A TEXT BOOK OF

FOUNDATION ENGINEERING

FOR
SEMESTER – VI

THIRD YEAR (T.Y.) B. TECH COURSE IN CIVIL ENGINEERING

Strictly According to New Revised Credit System Syllabus of Babasaheb Ambedkar Technological University (BATU), Lonere, (Dist. Raigad) Maharashtra, (w.e.f. June 2019-20)

Dr. SHRIKANT S. JAHAGIRDAR
M.E. (Envir.), Ph.D. (Civil, NITK, Surathkal)
Professor in Civil & Dean R & D.
N. K. Orchid College of Engineering & Technology
SOLAPUR.

Dr. SATISH B. MORE
M.E. (Geotech), Ph.D. (Civil, NITK, Surathkal)
Associate Professor, Civil Engg. Deptt.
N. K. Orchid College of Engineering & Technology
SOLAPUR.

N1102

FOUNDATION ENGINEERING (CIVIL, BATU)

ISBN : 978-93-89686-51-7

First Edition : **November 2019**

© : **Authors**

Published By :
NIRALI PRAKASHAN
Abhyudaya Pragati, 1312 Shivaji Nagar
Off J.M. Road, PUNE 411005
Tel : (020) 25512336/37/39
Email : niralipune@pragationline.com

➢ DISTRIBUTION CENTRES

PUNE

Nirali Prakashan (Local) : 119 Budhwar Peth, Jogeshwari Mandir Lane, Pune 411002, Maharashtra

Tel : (020) 2445 2044, Mobile : 9657703145, Email : niralilocal@pragationline.com

Nirali Prakashan (Outstation) : S. No. 28/27 Dhayari, Near Asian College, Dhayari, Pune 411041, Maharashtra

Tel : (020) 2469 0204, Fax : (020) 2469 0316, Mobile : 9657703143

Email : bookorder@pragationline.com

MUMBAI

Nirali Prakashan : 385 S.V.P. Road, Rasadhara Co-op. Hsg. Society Ltd., Girgaum, Mumbai 400004, Maharashtra

Tel : (022) 2385 6339 / 2386 9976, Fax : (022) 2386 9976, Mobile : 9320129587

Email : niralimumbai@pragationline.com

➢ DISTRIBUTION BRANCHES

JALGAON

Nirali Prakashan : 34 V. V. Golani Market, Navi Peth, Jalgaon 425001, Maharashtra

Tel : (0257) 222 0395, Mob : 94234 91860, Email : niralijalgaon@pragationline.com

KOLHAPUR

Nirali Prakashan : New Mahadvar Road, Kedar Plaza 1st Floor, Opp. IDBI Bank

Kolhapur 416012, Maharashtra. Mobile : 9850046155

Email : niralikolhapur@pragationline.com

NAGPUR

Nirali Prakashan : Above Maratha Mandir, Shop No 3, Second Floor,

Rani Jhanshi Square, Sitabuldi, Nagpur 440012, Maharashtra

Tel : (0712) 254 7129, Email : niralinagpur@pragationline.com

DELHI

Nirali Prakashan : 4593/15 Basement, Agarwal Lane, Ansari Road, Daryaganj

Near Times of India Building, New DelhiV 110002 Mobile : 8505972553

Email : niralidelhi@pragationline.com

BENGALURU

Nirali Prakashan : Maitri Ground Floor, Jaya Apartments, No. 99, 6th Cross, 6th Main,

Malleswaram, Bengaluru 560003, Karnataka

Mobile : 9449043034, Email : niralibangalore@pragationline.com

niralipune@pragationline.com | www.pragationline.com
Also find us on www.facebook.com/niralibooks

Dedicated to...

Our Beloved Students

....Authors

PREFACE

It gives us great pleasure to present the book **"Foundation Engineering"** for the students of **Semester VI Third Year (T.Y.) B. Tech. Course Civil Engineering of Dr. Babasaheb Ambedkar Technological University (BATU), Lonere, Dist. Raigad (Maharashtra).** This book is strictly as per the new revised syllabus 2019-20 Pattern, effective from the Academic Year July 2019-20.

In New Revised Syllabus, there will Class Assessment (CA) 20 Marks, Mid Sem. Exam. (MSE) 20 Marks and End Sem. Exam. (ESE) 60 Marks. End Sem. Exam. will be based on all Six units and each unit will carry 20 Marks.

The Theory Course will have 3 Credits.

The basic objective of this book is to bridge the gap between the vast contents of the reference books, written by the renowned International Authors and the concise requirements of Undergraduate Students. This book has been written in a comprehensive manner using Simple and Lucid language, keeping in mind students' requirements. The main emphasis has been given on exploring the basic concepts rather than merely the Information. Solved Examples and Exercises have been provided throughout the book and at the end of the Unit. Also, we have given **Model Question Papers** for practice at the end of book.

Our special thanks to our family members, students and all those who directly or indirectly supported us in this project.

We also take this opportunity to express our sincere thanks to Shri. Dineshbhai Furia, Shri. Jignesh Furia, Mrs. Nirali Verma, Shri. M. P. Munde and entire team of Nirali Prakashan, namely Mrs. Deepali Lachake (Co-ordinator), and her colleagues who really have taken keen interest and untiring efforts in publishing this text.

The advice and suggestions of our esteemed readers to improve the text are most welcome and will be highly appreciated.

Pune   **Authors**

SYLLABUS

Module I : Soil Exploration   (6 Lectures)

Introduction, General requirements to be satisfied for satisfactory performance of foundations, Soil exploration: Necessity, Planning, Exploration Methods, Soil Sampling Disturbed and undisturbed, Rock Drilling and Sampling, Core Barrels, Core Boxes, Core Recovery, Field Tests for Bearing Capacity evaluation, Test Procedure & Limitations

Module II : Bearing Capacity and Settlement   (7 Lectures)

Bearing Capacity Analysis - Failure Modes, Terzaghi's Analysis, Specialization of Terzaghi' s Equations, Skempton Values for Nc, Meyerhof's Analysis, I.S. Code Method of Bearing Capacity Evaluation, Effect of Water Table, Eccentricity of load, Safe Bearing Capacity and Allowable Bearing Pressure, Settlement Analysis: Immediate Settlement – Consolidation Settlement, Differential Settlement, Tolerable Settlement, Angular distortion

Module III : Foundations for Difficult Soils   (5 Lectures)

Guidelines for Weak and Compressible Soils, Expansive soil, Parameters of Expansive Soils, Collapsible Soils and Corrosive Soils, Causes of Moisture changes in Soils, Effects of Swelling on Buildings, Preventative Measures for Expansive Soils, Modification of Expansive Soils, Design of Foundation on Swelling Soils, Ground Improvement Methods: for general considerations, for Cohesive Soils, for Cohesionless Soils,

Module IV : Shallow Foundation   (5 Lectures)

Assumptions & Limitations of Rigid Design Analysis, Safe Bearing Pressure, Settlement of Footings, Design ofIsolated, Combined, Strap Footing (Rigid analysis), Raft Foundation (Elastic Analysis), I. S. Code of Practice for Design of Raft Foundation

Module V : Deep Foundation   (7 Lectures)

Pile Foundation: Classification, Pile Driving, Load Carrying Capacity of Piles, Single Pile Capacity, Dynamic Formulae, Static Formulae, Pile Load Tests, Penetration Tests, Negative skin Friction, Under Reamed Piles, Group Action of Piles, Caissons Foundations: Box, Pneumatic, Open Caissons, Forces, Grip Length, Well Sinking, Practical Difficulties And Remedial Measures Sheet Piles: Classification, Design of Cantilever Sheet Pile in Cohesionless and Cohesive soils. Design of Anchored Sheet Pile by Free Earth Support Method, Cellular Cofferdams: Types, Cell Fill Stability Considerations

Module VI : Stability of Slope   (6 Lectures)

Different Definitions of Factors of Safety, Types of Slope Failures, Stability of an Infinite Slope of Cohesionless Soils, Stability Analysis of an Infinite Slope of Cohesive Soils, Stability of Finite Slopes- Slip Circle Method, Semi Graphical and Graphical Methods, Friction Circle Method, Stability Number: Concept and its use

CONTENTS

✠ ✠ ✠

SOIL EXPLORATION

1.1 INTRODUCTION

Every building consists of two basic components which are super-structure and substructure or foundations. The super-structure is usually that part of the building which is above ground, and which serves the purpose of its intended use. The sub structure or foundations is the lower portion of the building, usually located below the ground level, which transmits the load of the super-structure to the sub-soil. Foundation is part of the structure which is in direct contact with the ground to which the loads are transmitted. The soil which is located immediately below the base of the foundation is called the sub-soil or foundation soil, while the lowermost portion of the foundation which is in direct contact with the sub-soil is called the footing.

1.1.1 Functions of Foundation

- **Reduction of Load Intensity :** Foundation distributes the loads of the super-structure, to a larger area so that the intensity of the load at its base does not exceed the safe bearing capacity of the sub-soil.

- **Even Distribution of Load:** Foundations distribute the none uniform load of the super-structure evenly to the sub-soil. For example, two columns carrying unequal loads can have a combined footing which may transmit the load to sub soil evenly with uniform soil pressure. Due to this, unequal or differential settlements are minimized.

- **Provision of Level Surface:** Foundation provide leveled and hard surface over which the super-structure can be built.

- **Lateral Stability:** The stability of the building, against sliding and overturning, due to horizontal forces (such as wind, earthquake etc.) is increased due to foundations.

- **Safety Against Undermining:** It provides structural safety against undermining or scouring due to burrowing animals and flood water.

- **Protection Against Soil Movements:** Special foundation measures prevents or minimizes the cracks in the super structure, due to expansion or contraction of the sub-soil because of moisture movement in some problematic soils.

1.1.2 Essential Requirements of a Good Foundation

Foundation should be constructed to satisfy the following requirements.

- The foundations shall be constructed to sustain the dead load and imposed loads and to transmit these to the sub-soil in such a way that pressure on it will not cause settlement which would impair the stability of the building or adjoining structures.

- Foundation base should be rigid so that differential settlements are minimized specially for the case when super imposed loads are not evenly distributed.

- Foundation should be taken sufficiently deep to ground, the building against damage or distress caused by swelling or shrinkage of the sub soil.

- Foundation should be so located that its performance may not be affected due to any unexpected future influence.

1.2 SOIL EXPLORATION

- Earthwork forms the largest activity of a Civil Engineer. It is well understood that irrespective of the type of civil engineering structure on earth –

  ➢ It has to be rested either in soil (e.g., foundations)

  ➢ Rested on soil (e.g., pavements) or

  ➢ The structure is itself constructed making use of soil (e.g., Earthen dams).

- This implies that a better knowledge of the spatial variation of the soils encountered is essential. Therefore, before construction of any civil engineering work a thorough investigation of the site is essential.

- Soil exploration/Site investigation is the exercise of undertaking a planned sequence of exploratory holes, with associated field and laboratory testing, in order to bring our understanding of the ground at a site, to an acceptable level of confidence for a particular project.

- Site investigation refers to the methodology of determining surface and subsurface features of the proposed area. Information on surface conditions **(surface exploration)** is necessary for planning the accessibility of site, for deciding the disposal of removed material (particularly in urban areas), for removal of surface water in water logged areas, for movement of construction equipments, and other factors that could affect construction procedures. This can be done by studying the various maps (topographical, contour, geological, aerial photographs etc.) and by reconnaissance of the area.

- Information on subsurface conditions **(subsurface exploration)** is more critical requirement in planning and designing the foundations of structures, dewatering systems, shoring or bracing of excavations, the materials of construction and site improvement methods can be done by various methods of exploration which is discussed below.

1.2.1 Necessity of Soil Exploration

- Site investigations constitute an essential and important engineering program which, while guiding in assessing the general suitability of the site for the proposed works, enables the engineer to prepare an adequate and economic design and to foresee and provide against difficulties that may arise during the construction phase. Site investigations are equally necessary in reporting upon the safety or causes of failures of existing works or in examining the suitability and availability of construction materials.

- **Objectives of Soil Exploration:** The information from soil investigations will enable a Civil engineer to plan, decide, design, and execute a construction project. Soil investigations are done to obtain the information that is useful for one or more of the following purposes

  ➢ To know the geological condition of rock and soil formation.

  ➢ To establish the groundwater levels and determine the properties of water.

  ➢ To select the type and depth of foundation for proposed structure

  ➢ To determine the bearing capacity of the site.

  ➢ To estimate the probable maximum and differential settlements.

  ➢ To predict the lateral earth pressure against retaining walls and abutments.

  ➢ To select suitable construction techniques

  ➢ To predict and to solve potential foundation problems

  ➢ To ascertain the suitability of the soil as a construction material.

  ➢ To determine soil properties required for design

  ➢ Establish procedures for soil improvement to suit design purpose

  ➢ To investigate the safety of existing structures and to suggest the remedial measures.

  ➢ To observe the soil performance after construction.

  ➢ To locate suitable transportation routes

  ➢ Selection of borrow areas for embankments.

1.2.2 Planning of Exploration Program

The actual planning of a subsurface exploration program includes some or all of the following steps:

1. **Assembly of All Available Information :**

- Dimensions, column spacing, type and use of the structure, basement requirements, any special architectural considerations of the proposed building, and tentative location on the proposed site. Foundation regulations in the local building code should be consulted for any special requirements. For bridges the soil engineer should have access to type and span lengths as well as pier loadings and their tentative location. This information will indicate any settlement limitations and can be used to estimate foundation loads.

2. **Reconnaissance of the Area :**

- This may be in the form of a field trip to the site, which can reveal information on the type and behavior of adjacent structures such as cracks, noticeable sags, and possibly sticking doors and windows. The type of local existing structures may influence to a considerable extent the exploration program and the best type of foundation for the proposed adjacent structure. Since nearby existing structures must be maintained in their "as is" condition, excavations or construction vibrations will have to be carefully controlled, and this can have considerable influence on the "type" of foundation that can be used.

- Erosion in existing cuts (or ditches) may also be observed, but this information may be of limited use in the foundation analysis of buildings. For highways, however, runoff patterns, as well as soil stratification to the depth of the erosion or cut, may be observed. Rock outcrops may give an indication of the presence or the depth of bedrock.

- The reconnaissance may also be in the form of a study of the various sources of information available, some of which include the following (maps, literature, etc.)

3. A Preliminary Site Investigation :

- In this phase a few borings (one to about four) are made or a test pit is opened to establish in a general manner the stratification, types of soil to be expected, and possibly the location of the groundwater table. If the initial borings indicate that the upper soil is loose or highly compressible, one or more borings should be taken to rock or competent strata.

- It is common at this stage to limit the recovery of good-quality samples to only three or four for laboratory testing. These tests, together with strength and settlement correlations using index properties such as liquid limit, plasticity index, and penetration test data as well as unconfined compression tests on disturbed samples recovered during penetration testing, are usually adequate for determining if the site is suitable.

4. A Detailed Site Investigation :

- Where the preliminary site investigation has established the feasibility and overall project economics, a more detailed exploration program is undertaken. The preliminary borings and data are used as a basis for locating additional borings, which should be confirmatory in nature, and determining the additional samples required. Note that if the soil is relatively uniformly stratified, a rather orderly spacing of borings at locations close to critical superstructure elements should be made.

- On occasion additional borings will be required to delineate zones of poor soil, rock outcrops, fills, and other areas that can influence the design and construction of the foundation. In the detailed program phase it is generally considered good practice to extend at least one boring to competent rock if the overlying soil is soft to medium stiff. This is particularly true if the structure is multiple-storied or requires settlement control

- The planning of an exploration program depends upon

  - ➢ The type and importance of the structure
  - ➢ The nature of the soil strata.
  - ➢ The depth, thickness, extent, and composition of each of the strata,
  - ➢ The depth of the rock,
  - ➢ The depth to the ground water table is important item sought to be determined by an exploration program.

The two important aspects of a boring program are

    (i) Spacing of borings and

    (ii) Depth of borings.

1.2.3 Exploration Methods

The methods available for soil exploration may be classified as follows:

    (I) Direct methods ... Test pits, trial pits or trenches

    (II) Semi-direct methods ... Borings

    (III) Indirect methods ... Soundings or penetration tests and geophysical methods

(I) Direct Method :

- **Test Pit :** In this method trench is excavated and the soil in the trench is visually examined and samples can be collected directly and can be used for determining strength and other properties of soil.

Fig. 1.1

- Applicable to all types of soils

- Provide for visual examination in their natural condition

- Disturbed and undisturbed samples can be conveniently obtained at different depths

- Depth of investigation: limited to 3 to 3.5 m.

Advantages :

- Cost effective

- Provide detailed information of stratigraphy

- Large quantities of disturbed soils are available for testing

- Large blocks of undisturbed samples can be carved out from the pits
- Field tests can be conducted at the bottom of the pits

Disadvantages :

- Depth limited to about 6m
- Deep pits uneconomical
- Excavation below groundwater and into rock difficult and costly
- Too many pits may scar site and require backfill soils.

Limitations :

- Undisturbed sampling is difficult
- Collapse in granular soils or below ground water table

(II) Semi-Direct Method :

- In this boreholes are drilled in order to obtain the samples of soil or rock at desired depth (which is called boring).
- The common method of advancing the boreholes are
  1. Auger boring
  2. Wash boring
  3. Percussion drilling
  4. Rotary drilling

1. **Auger Boring :** In this boreholes are drilled by using augers,which is a device that is useful for advancing a bore hole into the ground. Augers may be hand-operated or power-driven; the former are used for relatively small depths (less than 3 to 5 m), while the latter are used for greater depths. It is suitable in all soils above GWT but only in cohesive soil below GWT.

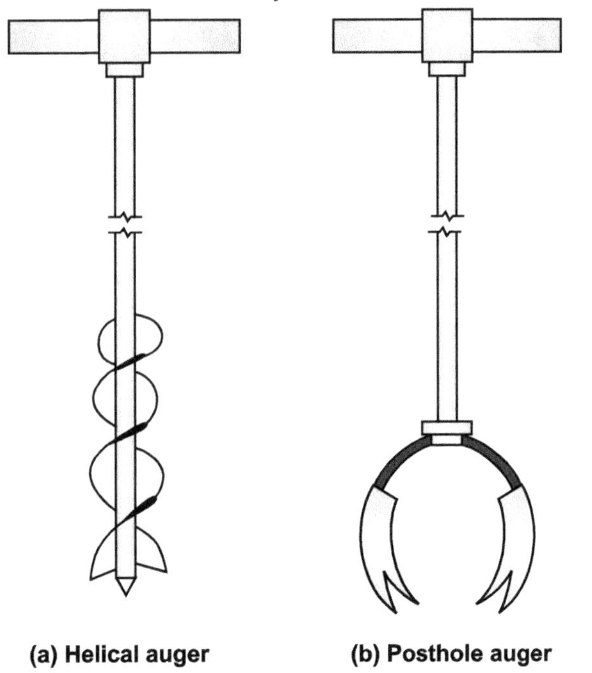

Fig. 1.2 : Hand tools

2. **Wash Boring :** The rotary wash boring method is generally the most appropriate method for use in soil formations below the groundwater level. In rotary wash borings, the sides of the borehole are supported either with casing or with the use of a drilling fluid

- Wash boring is commonly used for exploration below ground water table for which the auger method is unsuitable.
- This method may be used in all kinds of soils except those mixed with gravel and boulders.
- A casing pipe is pushed in and driven with a drop weight.
- A hollow drill bit is screwed to a hollow drill rod connected to a rope passing over a pulley and supported by a tripod.
- Water jet under pressure is forced through the rod and the bit into the hole. This loosens the soil at the lower end and forces the soil-water suspension upwards along the annular surface between the rod and the side of the hole.
- This suspension is led to a settling tank where the soil particles settle while the water overflows into a sump. The water collected in the sump is used for circulation again
- The soil particles collected represent a very disturbed sample and is not very useful for the evaluation of the engineering properties.
- Wash borings are primarily used for advancing bore holes; whenever a soil sample is required, the chopping bit is to be replaced by a sampler.
- The change of the rate of progress and change of colour of wash water indicate changes in soil strata

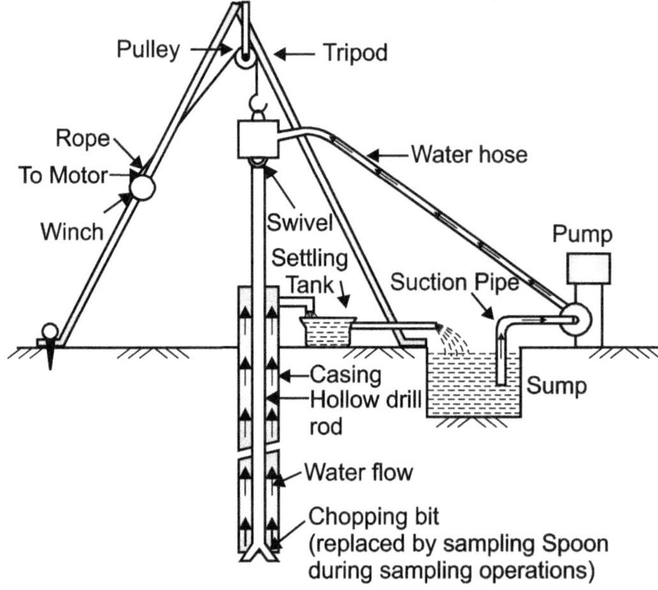

Fig. 1.3 : Set-up for wash boring

- Where drill casing is used, the boring is advanced sequentially by:
  - Driving the casing to the desired sample depth,
  - Cleaning out the hole to the bottom of the casing, and
  - Inserting the sampling device and obtaining the sample from below the bottom of the casing.

3. **Percussion Boring:** Grinding the soil by repeated lifting and dropping of heavy chisels or drilling bits. Water is added to form slurry of cuttings. Slurry removed by bailers or pumps. In general, a machine used to drill holes is called a drill rig (generally power driven, but may be hand driven). A winch is provided to raise and lower the drilling tools into the hole. The slurry of pulverized material is bailed out at intervals. The method is suitable for advancing a hole in all types of soils, boulders and rock. The formation, however, get disturbed by the impact.

4. **Rotary Drilling:** Rotary Boring or rotary drilling is a very fast method of advancing hole in both rocks and soils. A drill bit, fixed to the lower end of the drill rods, is rotated by a suitable chuck, and is always kept in firm contact with the bottom of the hole. A drilling mud, usually a water solution of bentonite, with or without other admixtures, is continuously forced down to the hollow drill rods. The mud returning upwards brings the cutting to the surface. The method is also known as mud rotary drilling and the hole usually requires no casing.

Fig. 1.4 : Drilling bit for percussion drilling

(III) Indirect Method : In these methods properties of soil are determined indirectly by measuring some other parameter of soil which is then correlated with the properties of soil. Following are the methods in this category
  - Sounding method (SPT/SCPT/DCPT)
  - Geo-physical methods (Electrical resistivity/seismic refraction)

- **Sounding Method :** In these methods probe is inserted in the ground (statically or dynamically) and resistance offered by the soil against its penetration is recorded. following are the various sounding methods.
  - Standard Penetration Test (SPT).
  - Static Cone Penetration Test (SCPT).
  - Dynamic Cone Penetration Test (DCPT).

1.2.4 Soil Sampling

- Basic aim of soil exploration is to determine soil properties for which soil is collected from the site of exploration and is then tested in the laboratory. The process of collecting the soil from the field is called soil sampling, soil collected is called soil sample and the instrument used for collecting the soil sample is called soil sampler.

- **Need for Sampling :** Sampling is carried out in order that soil and rock description, and laboratory testing can be carried out.

  Laboratory tests typically consist of:
  - Index tests (for example, specific gravity, water content).
  - Classification tests (for example, Atterberg limit tests on clays);
  - Tests to determine engineering design parameters (for example Strength, compressibility, and permeability).

Factors to be Considered While Sampling of Soil

- Samples should be representative of the ground from which they are taken.

- They should be large enough to contain representative particle sizes, fabric, and fissuring and fracturing.

- They should be taken in such a way that they have not lost fractions of the situ soil (for example, coarse or fine particles) and, where strength and compressibility tests are planned they should be subject to as little disturbance as possible.

- **Types of Soil Samples:**

Soil samples are classified as below

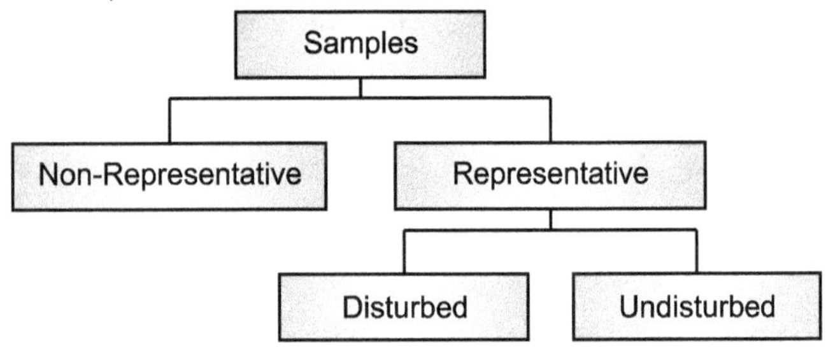

Fig. 1.5

1. **Non-Representative Samples :** Non-Representative soil samples are those in which neither the in-situ soil structure, moisture content nor the soil particles are preserved.

   - They are not representative.
   - They cannot be used for any tests as the soil particles either gets mixed up or some particles may be lost.

   e.g., Samples that are obtained through wash boring or percussion drilling.

2. **Representative Samples :** Representative soil samples are those in which constituent minerals are retained but structure of soil gets disturbed. There are two types if representative soil samples
   (i) Disturbed soil sample,
   (ii) Undisturbed soil sample

   (i) Disturbed Soil Samples : Disturbed soil samples are those in which the in-situ soil structure and moisture content are lost, but the soil particles are intact.
   - They are representative.
   - They can be used for grain size analysis, liquid and plastic limit, specific gravity, compaction tests, moisture content, organic content determination and soil classification test performed in the lab

   e.g., obtained through cuttings while auguring, grab, split spoon (SPT), etc

   (ii) Undisturbed Soil Samples : Undisturbed soil samples are those in which the in-situ soil structure and moisture content are preserved.
   - They are representative and also intact.
   - These are used for **consolidation, permeability or shear strengths test (Engineering properties)**.
   - More complex jobs or where **clay** exist
   - In **sand** is very difficult to obtain undisturbed sample.

- Obtained by using Shelby tube (thin wall), piston sampler, surface (box), vacuum, freezing, etc.

- **Causes of Soil Disturbances**
  - Friction between the soil and the sampling tube.
  - The wall thickness of the sampling tube.
  - The sharpness of the cutting edge.
  - Care and handling during transportation of the sample tube.

To minimize friction the sampling tube should be pushed instead of driven into the ground.

Sampling tube that are in common use have been designed to minimize sampling disturbances.

- **Design Features Affecting the Sample Disturbance**
  - Area ratio
  - Inside Clearance
  - Outside Clearance
  - Recovery Ratio
  - Inside wall friction
  - Design of non-return value
  - Method of applying force
  - Sizes of sampling tubes

Area Ratio: It is defined as the ratio of c/s area of cutting edge of sampler to the c/s area of sample

$$A_r = \frac{D_2^2 - D_1^2}{D_1^2}$$

- For obtaining good quality undisturbed samples, the area ratio should be less than or equal to 10%.
- It may be high as 110% for thick wall sampler like split spoon sampler and may be as low as 6 to 9% for thin wall samples like Shelby tube.

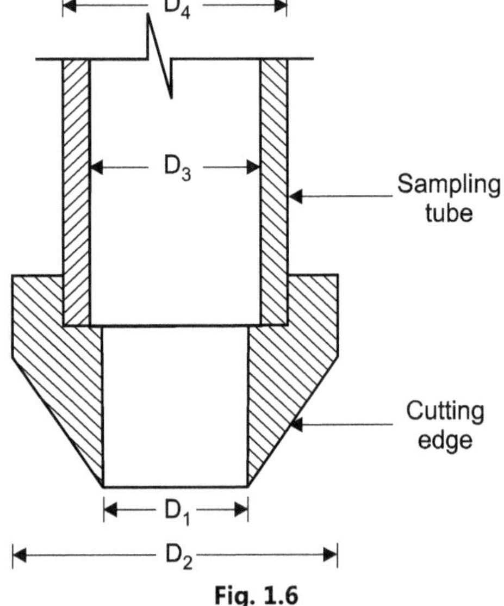

Fig. 1.6

Inside Clearance: It is defined as ratio of difference between inner diameter of cutting edge and sampling tube to that of inner diameter of sampling tube

$$C_I = \frac{D_3 - D_1}{D_1} \times 100$$

- The inside clearance allows **elastic expansion** of the sample when it enters the sampling tube.
- It helps in reducing the **frictional drag** on the sample, and also helps **to retain the core.**
- For an undisturbed sample, the inside clearance should be between **0.5 and 3%.**

Outside Clearance: It is defined as the ratio of difference of outer diameter of the cutting edge and sampling tube to that of outer diameter of the sampling tube.

$$C_o = \frac{D_2 - D_4}{D_4} \times 100$$

- Outside clearance facilitates the **withdrawal** of the sample from the ground.
- For **reducing the driving force**, the outside clearance should be as small as possible.
- Normally, it lies between zero and 2%.
- C_O Should not be more than C_I

Recovery Ratio: It is defined as the ratio of the actual length of sample recovered from the sampling tube to the depth of penetration of sampling tube in to the ground.

$$R_r = \frac{L}{H} \times 100$$

$$R_r = 96 - 98 \text{ \% for getting a satisfactory undisturbed sample}$$

Inside Wall Friction

- The friction on the inside wall of the sampling tube causes disturbances of the sample.
- Therefore, the inside surface of the sampler should be as smooth as possible.
- It is usually smeared with oil before use to reduce friction

Design of Non-Return Value

- The non – return value provided on the sampler should be of proper design.
- It should have an orifice of large area to allow air, water or slurry to escape quickly when the sampler is driven.
- It should close when the sample is withdrawn

Method of Applying Force

- The degree of disturbance depends upon the method of applying force during sampling and depends upon the rate of penetration of the sample.
- For obtaining **undisturbed samples, the sampler should be pushed and not driven.**

1.3 DEPTH OF EXPLORATION

The degree of variation of the sub-surface data in the horizontal and vertical directions decides the depth of exploration required at a particular site. It is impossible to fix the number, disposition and depth of borings without making a few preliminary borings or soundings at the site.

In general, exploration should be carried out to a depth upto which the increase in pressure due to structural loading is likely to cause significant settlement or sheer failure. Such depth is known as the 'Significant Depth'.

The significant depth depends upon :

- Type of structure.
- Weight of structure.
- Size of structure.
- Shape of structure.
- Disposition of loaded areas.
- Soil profile and its properties.

The significant depth is taken as the depth at which the vertical stress is 20% of load intensity.

Guide Rules for Depth of Exploration :

1. **Square Footing :** Depth of exploration should be about 1.5 times the width of the footings.

2. **Strip Footing :** Three times the width of footing.

3. **Adjacent Footings with Clear Spacing Less Than the Twice the Width :** The depth of boring should be minimum 1.5 times the width of entire loaded area.

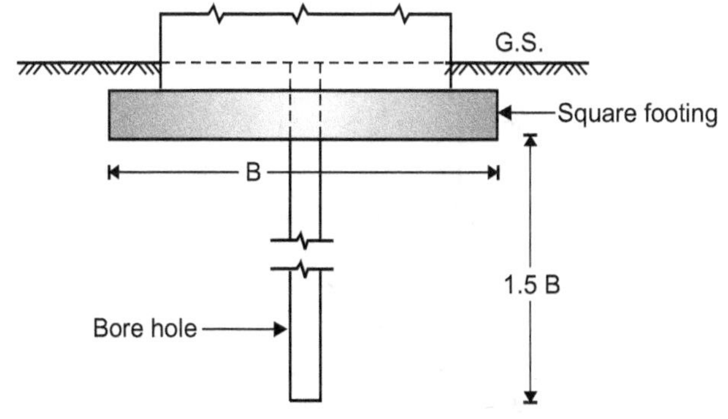

Fig. 1.7 : Depth of exploration

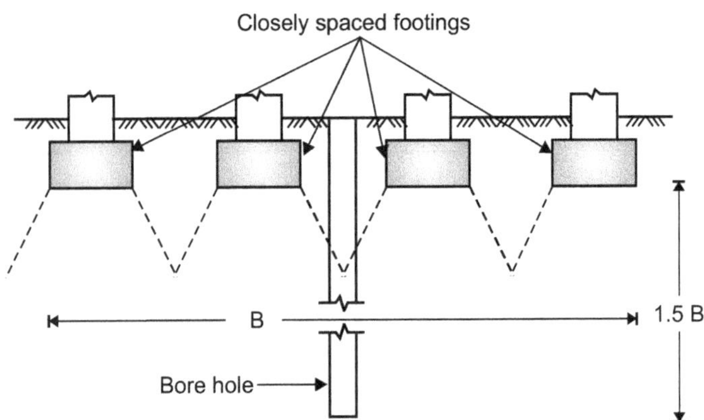

Fig. 1.8 : Depth of exploration for closely-spaced footings

4. **Pile Foundation :** Depth of exploration in case of pile foundations is 10 to 30 m or more or atleast 1.5 times the width of the pile group.

5. **Friction Piles :** In case of friction piles depth of exploration is taken 1.5 times the width of the pile group measured from the lower third point.

6. **Base of Retaining Wall :** One and a half times the base width or one and a half times the exposed height of face of wall, whichever is greater.

7. **Floating Basements :** Depth of explorations may be taken as equal to depth of construction.

8. **Multi-Storey Buildings :** In case of multi-storey buildings depth of exploration can be worked out from the following formula.

$$D = C(S)0.7$$

Where,　　　D = Depth of exploration

　　　　　　C = Constant

　　　　　　S = Number of storeys

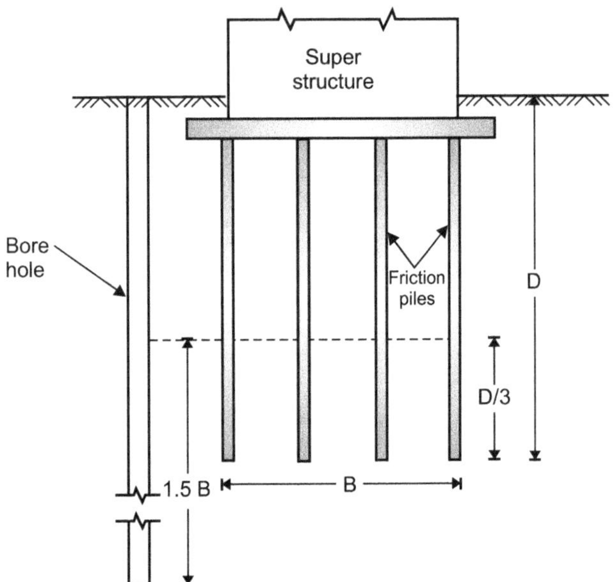

Fig. 1.9: Depth of exploration for friction piles

Table 1.1

Value of C	Types of structure
3.0	For light steel and narrow concrete buildings.
6.0	For heavy steel and wide concrete buildings.

9. **Foundations on Rock :** The minimum depth of core boaring into the bed rock should be 3 m to establish it as a rock.

10. **Dams :** For earth dams, depth of exploration is 1.5 times bottom width of the dam. For concrete dam depth of exploration is two times the height of dam from stream bed.

11. **Road-Cuts :** The depth of exploration is taken equal to the width of the cut.

12. **Road-Fills :** The depth of exploration is equal to height of the fill or the minimum depth of boring, which is 2 m, whichever is greater.

1.4 NUMBER OF EXPLORATION HOLES

The lateral extent of explorations and the spacing of bore holes should be such as to reveal any major changes in thickness, depth or properties of the strata affected by the works and the adjacent surroundings.

Following criteria shall be followed :

1. **Small and Less Important Buildings :** For these buildings even one bore hole or a trial pit in the centre is sufficient.

2. **Compact Buildings :** For compact buildings which covers area of about 0.4 hectares there should be atleast 5 bore holes, one at the centre and four near corners.

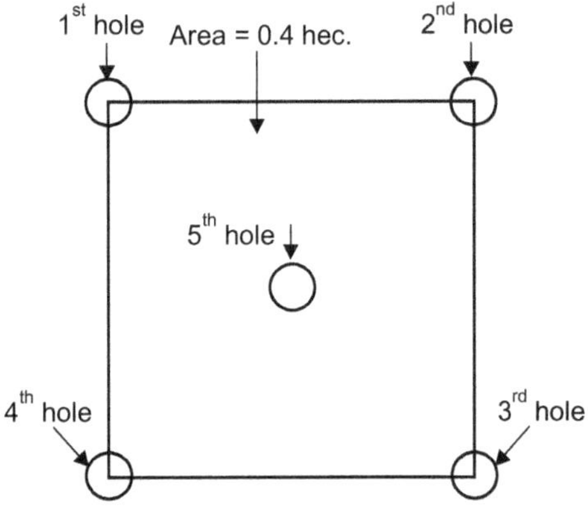

Fig. 1.10 : Lateral extent of explorations for compact buildings

3. **Large and Multi-Storyed Buildings :** The bore holes should be drilled at important places and at all corners. In this case, spacing of bore holes is generally kept between 10 m to 30 m, depending upon the variation in the sub-surface conditions and loading.

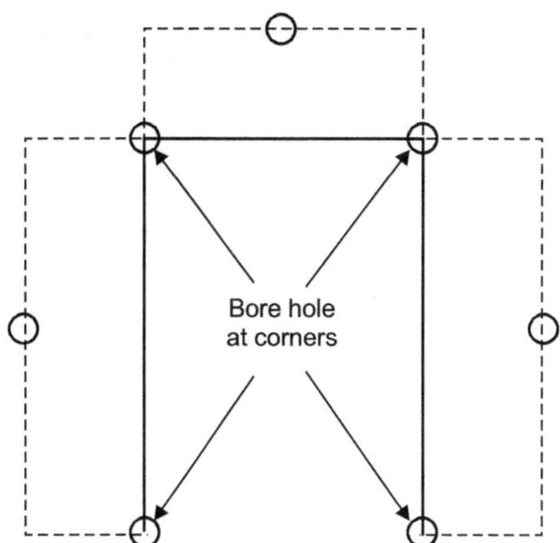

Fig. 1.11 : Lateral depth of explorations for large and multi-storyed buildings

4. **Highways :** For highways, sub-surface explorations are carried out along the centre line or along the proposed ditch line. In this case, spacing of bore holes generally varies between 150 to 300 m. If sub-strata is irregular, the spacing may be reduced to minimum 30 m.

5. **Dams :** Generally, the spacing of bore holes varies between 40 to 80 m. In addition to bore holes along the top line of upstream face of the dam, a few more widely scattered holes are also required in the bottom of the reservoir on the upstream side of the dam.

1.5 ROCK CORE DRILLING AND SAMPLING (NON-DESTRUCTIVE ROCK CORE DRILLING)

Rock core drilling is the process of recovering cylindrical samples (cores) of bedrock. A core barrel equipped with a drill bit is attached to the bottom of the drill rods, and as the rods and core barrel are rotated and advanced, the core is collected in the core barrel. Drilling fluid, which typically consists of water but can also be a variety of slurries, is pumped through the drilling rods and core barrel during advancement to flush cuttings from the boring and to lubricate and cool the drill bit.

Bedrock sampling is required for most subsurface explorations, particularly subsurface explorations for structures and roadway rock cuts and for project sites where rock is at a shallow depth. When bedrock sampling is required, rock core drilling must be performed. Rock core sampling is performed to:

- Determine the elevation of the top of bedrock

- Obtain rock core samples for visual inspection and laboratory testing
- Identify the bedrock type
- Describe bedrock color, hardness, weathering, bedding and discontinuity spacing, dip magnitude, and Rock Quality Designation (RQD).

During rock core sampling, subsurface features encountered that could influence the determination of a proper foundation may include voids, mines, or other zones of weakness such as soil or clay infillings, discontinuities, and shear zones/faults. These features are often identified not only in the recovered rock core samples but also during changes in drilling such as:

- **Rate of Drilling :** Increased or decreased drilling rate, which include tool drops or rapid rates of advancement as these could be indicators of voids, soft zones, clay soil seams, or highly weathered rock. Decreased drilling rates or lack of advancement could indicate very hard rock conditions. Blocking off or plugging of the core barrel can indicate highly fractured rock conditions or presence of clay seams.

- **Drill Water Conditions :** Changes in the amount of drill water return to the surface or loss of drill water return entirely can indicate the presence of voids, open discontinuities, faults, weathered zones, and soil seams. Drill water return color should be documented for each core run as this can be an indicator to the degree of rock weathering.

- **Rock Core Sample Observations and Handling :** If rock core losses are observed documentation should identify and provide an interpretation for the core loss particularly if the losses are thought to represent conditions different form the core recovered. Measurements and observations such as core loss, core recovery, and RQD should be measured while the core remains in the split inner tube, and not after core segments are fitted together and placed in the core box. Rock core samples should be oriented so that bedding or foliation is observed at its maximum dip.

1. **Core Barrel :**

- A hollow cylinder attached to a specially designed bit and which is used to obtain and to preserve a continuous section, or core, of the rocks penetrated in drilling.

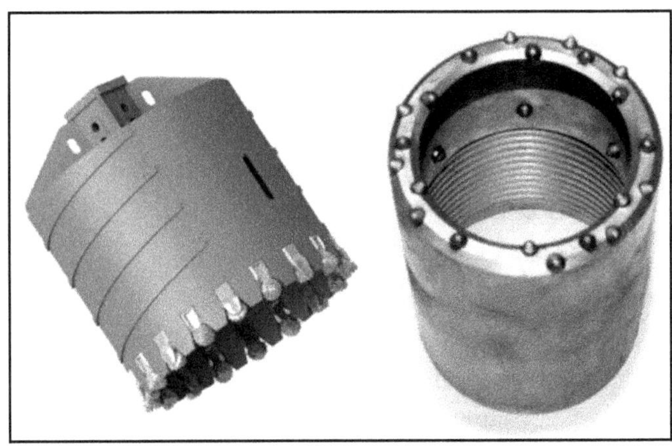

Fig. 1.12

2. Core Boxes:

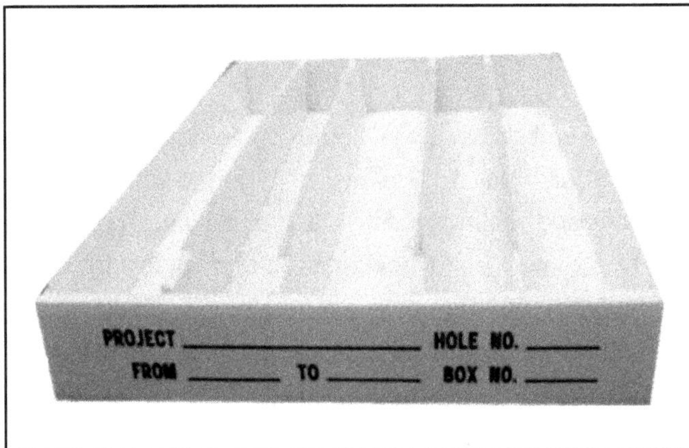

Fig. 1.13

- These are the box which is used for placing and preserving samples after they were taken out from the core barrels and is further used for investigation to find its quality and strength.

1.6 FIELD TESTS FOR BEARING CAPACITY

The test which is conducted at the site of construction before constructing the foundation is called field test.

Advantages of Field Tests :

- Sampling not required.
- Soil disturbance minimum.
- Large soil mass is tested which is not possible in the lab.

Disadvantages of Field Tests :

- Laborious.
- Time consuming.
- Cannot control initial state of stress during test.
- Many times stress induced during the testing are horizontal while building loads are vertical.
- Heavy equipment to be carried to field.
- Short duration behavior.
- Many times results are empirical.

Bearing capacity of soil in the field can be evaluated by using following two commonly used methods :

1. Plate load test and
2. Standard penetration test

1.6.1 Plate Load Test

Object: Field test to determine ultimate bearing capacity of soil and the probable settlement under a given loading.

Specifications: [IS 1888-1962]

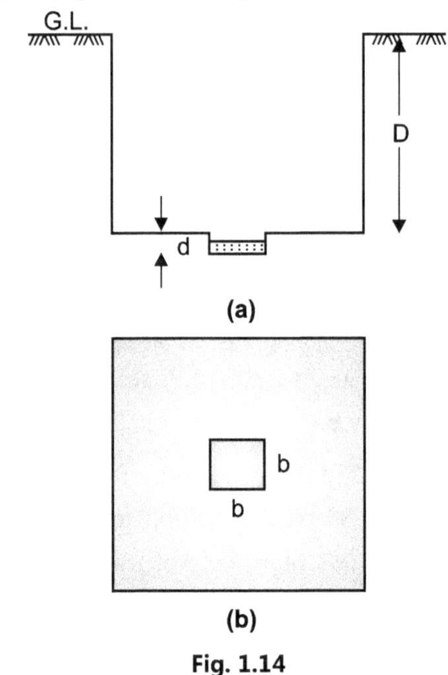

Fig. 1.14

Bearing Plate :

- **Shape :** (square or circular), size (300mm to 750mm),
- **Thickness:** decided based on bending stress consideration however it should not be less than 25mm.

Pit:

- **Size :** Five times the size of plate,
- **Depth :** Should satisfy $\dfrac{D}{B} = \dfrac{d}{b}$

Procedure : The experimental arrangement is as shown.

1. To start with seating load (7kPa) is applied (to Compensate unevenness of ground), which is released before starting actual test and dial gauges are set to read zero.

2. Loads are applied in cumulative load increments of not more than minimum of following values

   - 20% of expected safe bearing capacity,
   - 10% of ultimate bearing capacity
   - Twice the design load.

3. Each load is maintained constant until the settlement rate is less than 0.002 mm/hr.

4. Settlements are recorded at following time interval of 1, 2, 5, 10, 20, 40min, 1hr, 2 hr, 3 hr ... 24 hrs.

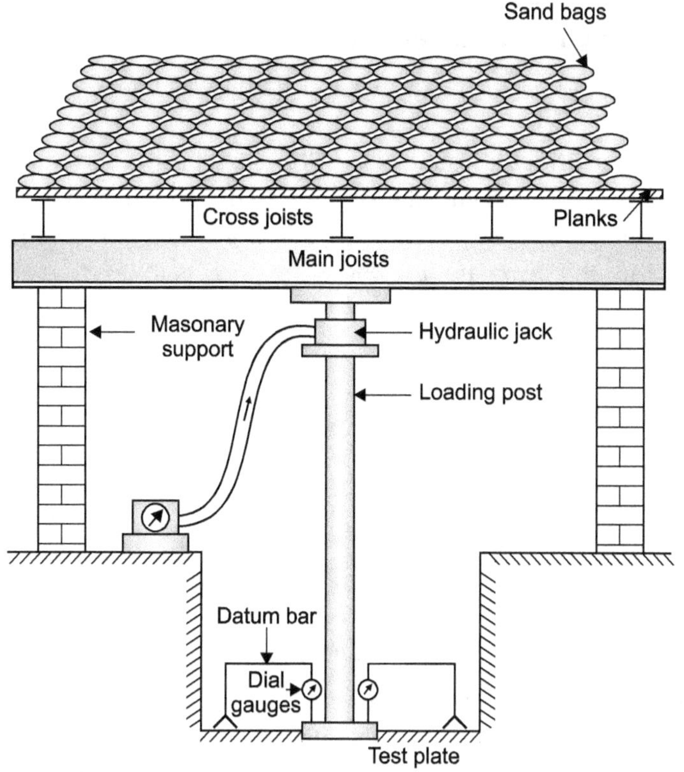

Fig. 1.15 : Plate load test set up

5. Apply the next increment and repeat the procedure.

6. Testing is continued until one of following stage is reached

   - Applied pressure exceeds 3 times allowable pressure

   - Total settlement exceeds 10% width of plate

Interpretation:

- **Ultimate Bearing Capacity of Plate (q_{up}) :** It is the ordinate corresponding to the point where graph changes its direction suddenly; intersection of tangent drawn to initial and final portion of the curve, load corresponding to settlement of 20% width of plate and if the graph consist of vertical q_u (p) is obtained from the graph.

- **Ultimate Bearing Capacity of Footing (q_{uf}) :**

   ➢ For clayey soil $q_{uf} = q_{up}$

   ➢ For sandy soil $q_{uf} = q_{up} \dfrac{B}{b}$

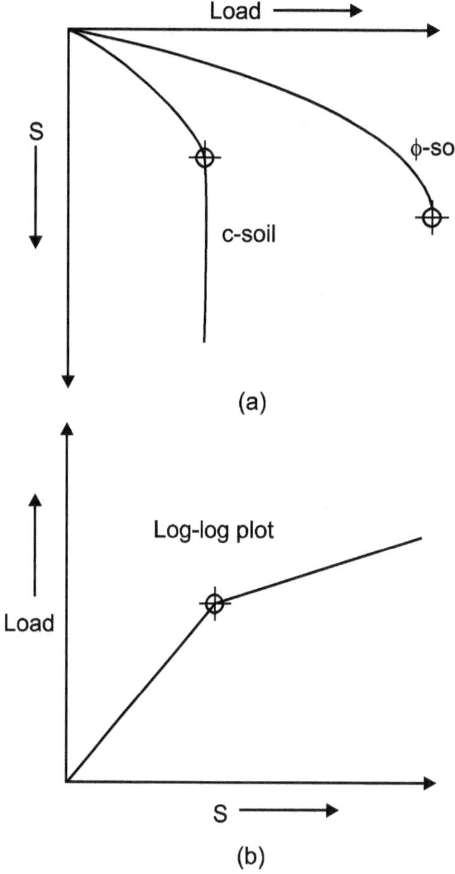

Fig. 1.16

- **Settlement of Footing:** If for a given pressure settlement of plate is S_p, then settlement of footing for same pressure is given by

   ➢ Clayey soil

   $$S_f = S_p \frac{B}{b}$$

   ➢ Sandy soil

   $$S_f = S_p \left[\frac{B(b + 0.3)}{b (B + 0.3)} \right]^2$$

Note : In above equations b is width of plate and B is width of footing **both need to be taken in meters.**

- **Safe Settlement Pressure: (Taylors Method)**

   $$\frac{q}{S} = C_1 \left[1 + 2 \frac{D}{B} \right] + \frac{C_2}{B}$$

   Where, S : Permissible settlement,

   q : safe settlement pressure

   Band D : Width and depth of footing

   C_1 and C_2 : Soil parameters to be obtained from two plate load test

- **Effect of Depth on Settlement:**

  When depth of embedment of footing is increased, the settlement of footing will be decreased. If S_1 and S_2 are settlement of footing (size B) at depth D_1 and D_2. These settlements are related by following relation

$$\frac{S_1}{S_2} = \sqrt{\frac{1 + 2\left(\frac{D_2}{B}\right)}{1 + 2\left(\frac{D_1}{B}\right)}}$$

The load settlement curve should pass through the origin and in case it does not, the "zero correction" is determined by extending the earliest straight line portion to intersect the vertical axis. The intercept on settlement axis at zero load is "**zero correction**". Corrected load settlement curve is obtained by substituting the correction from the recorded settlement.

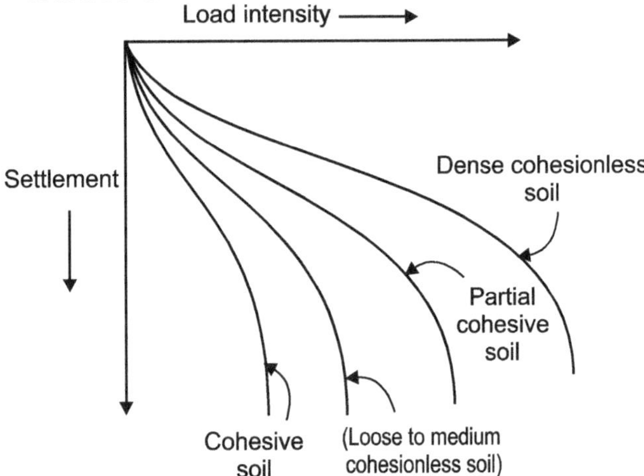

Fig. 1.17 : Typical load settlement curve

Advantages of Plate Load Test :
- It provides the allowable bearing pressure at the location considering both shear failure and settlement.
- Being a field test, there is no requirement of extracting soil samples.
- The loading techniques and other arrangements for field testing are identical to the actual conditions in the field.
- It is a fast method of estimating ABP and P – Δ behavior of ground.

Disadvantages of Plate Load Test :
- The test results reflect the behavior of soil below the plate (for a distance of ~2Bp), not that of actual footing which is generally very large.
- It is essentially a short duration test. Hence, it does not reflect the long term consolidation settlement of clayey soil.

- Size effect is pronounced in granular soil. Correction for size effect is essential in such soils.
- It is a cumbersome procedure to carry equipment, apply huge load and carry out testing for several days in the tough field environment.

Limitations of Plate Load Test:

Results obtained by PLT may not be representative due to following factors :

- **Size Effect :** Results of the PLT test reflect the strength and settlement characteristics of soil within the pressure bulb of the test plate, which may not extend to the zone up to which pressure bulb of footing extends.

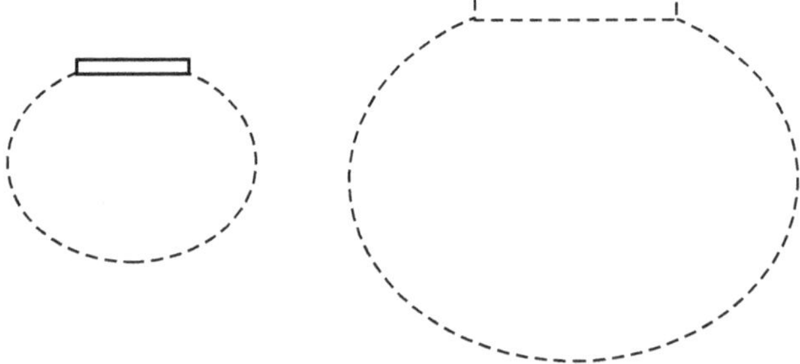

Fig. 1.18

- **Scale Effect :** Bearing capacity of cohesive soil is independent of size of footing, and that of cohesionless soil depends on size of footing whereas settlement in both types of soil is function of size of footing. Thus bearing capacity and settlement of actual foundation will differ from those obtained due to small test plate.

- **Time Effect :** PLT is short duration test, thus for clayey soil it does not give ultimate settlement.

- **Interpretation of Failure Load :** The failure load in PLT is not well defined except in case of general shear failure, so the load obtained in other types of failure will differ.

- **Water Table Effect:** The position of water table lying below the test level has no influence on bearing capacity of plate however bearing capacity of actual footing get effected by water table position when it lies above the significant depth level.

- **Site and Season:** It is difficult to obtain results that represent the whole site and the whole season. In summer it gives good results compared to that in wet season.

- **Shape of Foundation:** Test ignores the effect of shape of footing on bearing capacity and settlement. As such results of PLT test using square plate cannot be applied to circular, rectangular or strip footing.

1.6.2 Standard Penetration Test (IS 2131 – 1981)

Procedure

1. The borehole is advanced to the required depth and the bottom cleaned.
2. The split-spoon sampler, attached to standard drill rods of required length is lowered into the borehole and rested at the bottom
3. The split-spoon sampler is driven into the soil for a distance of 450mm by blows of a drop hammer (monkey) of 65 kg falling vertically and freely from a height of 750 mm.
4. The number of blows required to penetrate every 150 mm is recorded while driving the sampler.
5. The number of blows required for the last 300 mm of penetration is added together and recorded as the N value at that particular depth of the borehole.
6. The number of blows required to effect the first 150mm of penetration, called the seating drive, is disregarded.

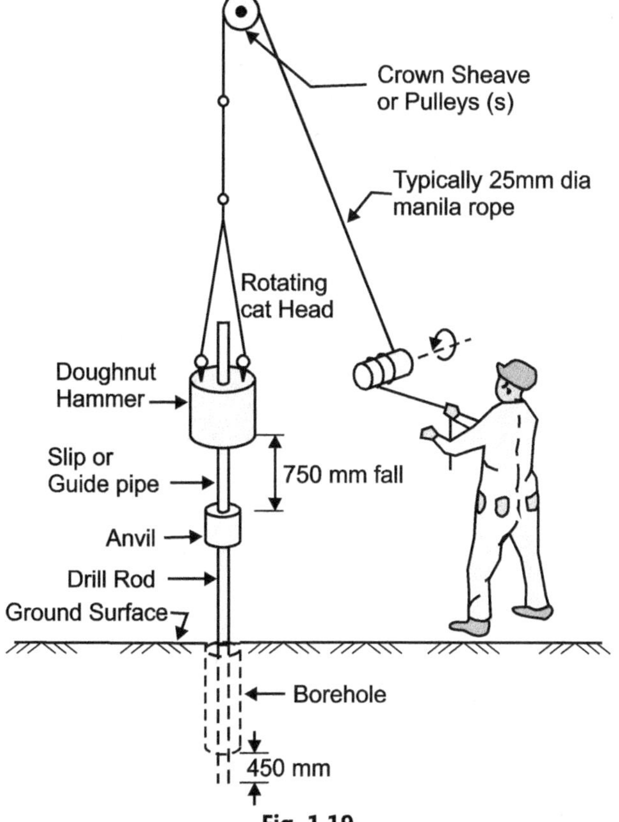

Fig. 1.19

- The split-spoon sampler is then withdrawn and is detached from the drill rods. The split-barrel is disconnected from the cutting shoe and the coupling. The soil sample collected inside the split barrel is carefully collected so as to preserve the natural moisture content and transported to the laboratory for tests. Sometimes, a thin liner is inserted within the split-barrel so that at the end of the SPT, the liner containing the soil sample is sealed with molten wax at both its ends before it is taken away to the laboratory.

- The SPT is carried out at every 0.75 m vertical intervals in a borehole. This can be increased to 1.50 m if the depth of borehole is large. Due to the presence of boulders or rocks, it may not be possible to drive the sampler to a distance of 450 mm. In such a case, the N value can be recorded for the first 300 mm penetration.

- The boring log shows refusal and the test is halted if 50 blows are required for any 150mm penetration 100 blows are required for 300m penetration 10 successive blows produce no advance.

Experimental Arrangement for the Test

Precautions

- The drill rods should be of standard specification and should not be in bent condition.

- The split spoon sampler must be in good condition and the cutting shoe must be free from wear and tear. The drop hammer must be of the right weight and the fall should be free, frictionless and vertical. The height of fall must be exactly 750 mm. Any change from this will seriously affect the N value.

- The bottom of the borehole must be properly cleaned before the test is carried out. If this is not done, the test gets carried out in the loose, disturbed soil and not in the undisturbed soil. When a casing is used in borehole, it should be ensured that the casing is driven just short of the level at which the SPT is to be carried out. Otherwise, the test gets carried out in a soil plug enclosed at the bottom of the casing.

- When the test is carried out in a sandy soil below the water table, it must be ensured that the water level in the borehole is always maintained slightly above the ground water level. If the water level in the borehole is lower than the ground water level, 'quick' condition may develop in the soil and very low N values may be recorded.

- In spite of all these imperfections, SPT is still extensively used because the test is simple and relatively economical. It is the only test that provides representative soil samples both for visual inspection in the field and for natural moisture content and classification tests in the laboratory. SPT values obtained in the field for sand have to be corrected before they are used in empirical correlations and design charts. IS: 2131-1981 recommends that the field value of N be corrected for two effects, namely,

1. Effect of overburden pressure, and
2. Effect of dilatancy.

1. Correction for Overburden Pressure:

- Several investigators have found that the penetration resistance or the N value in a granular soil is influenced by the overburden pressure.
- Of two granular soils possessing the same relative density but having different confining pressures, the one with a higher confining pressure gives a higher N value.
- Since the confining pressure (which is directly proportional to the overburden pressure) increases with depth, the N values at shallow depths are underestimated and the N values at larger depths are overestimated. To allow for this, N values recorded from field tests at different effective overburden pressures are corrected to a standard effective overburden pressure.

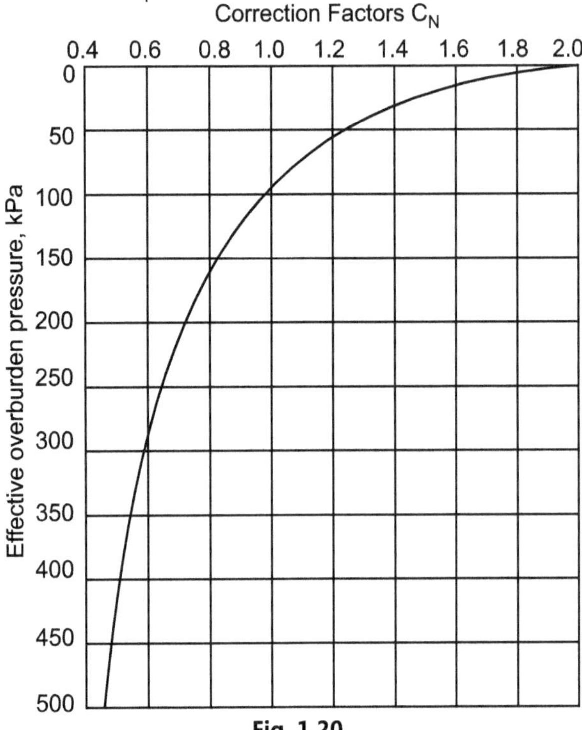

Fig. 1.20

$$N' = C_N N_O$$

where N_O : Observed value of blow count

C_N : Normalized correction factor

N' : Blow count corrected for Overburden

The value of normalized correction factor C_N can be obtained by using chart or by using the formula as given below :

$$C_N = 0.77 \log \frac{2000}{\sigma_o} \quad \text{... (for } \sigma_o \geq 25\text{kPa)}$$

where σ_o : effective Overburden pressure at the test level

at σ_o = 100kPa Correction factor =1,

at ground surface correction factor = 2

at σ_o = 10kPa correction factor = 1.8,

at σ_o = 20kPa correction factor = 1.6

at σ_o = 25kPa correction factor = 1.5,

at σ_o = 450kPa correction factor = 0.5

2. Dilatancy Correction :

- The value of N_o corrected for overburden need to be corrected for dilatancy. If the subsoil consists of fine sand and silt below the water table. (Sand and silt below water table offer higher resistance to driving due to development of excess pore pressure, which could not be dissipated immediately)

Note : This correction is applied only when N_1 is more than 15

$$N_c = 15 + 0.5 (N' - 15)$$

Correlation of blow count value with various soil properties are as below

Table 1.2

For Sandy Soil			
Nc	Relative Density	φ	
< 4	15%	28	Very loose
4-10	15% - 35%	30	Loose
10-20	35% - 50%	32	Moderately dense
20-30	50% - 65%	36	Medium dense
30-50	85% - 99%	42	Dense
> 50	100%	45	Very dense

Table 1.3

For Clayey Soil		
Nc	Consistency	UCS kPa
0-2	Very soft	< 25
2-4	Soft	25-50
4-8	Firm to medium	50-100
8-16	Stiff	100 -200
16-32	Very stiff	200 - 400
>32	Hard	> 400

Soil Modulus:

for sandy soil E = 500(N+15) kPa

clayey soil E = 300(N+5) kPa

The values of C and φ were obtained from value of Nc and then by using Terzaghi equation bearing capacity can be calculated.

- For strip footing :

$$q_{nu} = [0.471BN^2W_r + 0.785(100 + N^2)dW_q] \text{ kPa}$$

- For square or circular footing

$$q_{nu} = [0.314BN^2W_r + 0.943(100 + N^2)dW_q] \text{ kPa}$$

where   B : Is width of footing,

W_q and W_r : water table correction factors

Advantages of Standard Penetration Test :

- Relatively quick & simple to perform.
- Equipment & expertise for test is widely available.
- Provides representative soil sample.
- Provides useful index for relative strength & compressibility of soil.
- Able to penetrate dense & stiff layers.
- Results reflect soil density, fabric, stress strain behavior.
- Numerous case histories available.

Disadvantages of Standard Penetration Test :

- Requires preparation of bore hole.
- Dynamic effort is related to mostly static performance.
- SPT is abused, standards regarding energy are not uniform.
- If hard stone is encountered, difficult to obtain reliable result.
- Test procedure is tedious and requires heavy equipment.
- Not possible to obtain properties continuously with depth.

SOLVED EXAMPLES

Example 1.1: *Compute the area ratio of a thin walled tube sampler having an external diameter of 6 cm and a wall thickness 2.25 mm. Do you recommend the sampler for obtaining undisturbed samples? Why?*

Solution:

(i)   Area ratio is given by,

$$A_r = \frac{D_2^2 - D_1^2}{D_1^2}$$

Inner diameter (D_1) = Outer diameter (D_2)

D_1 = 60 mm – (2.25 × 2)

D_1 = 55.50 mm

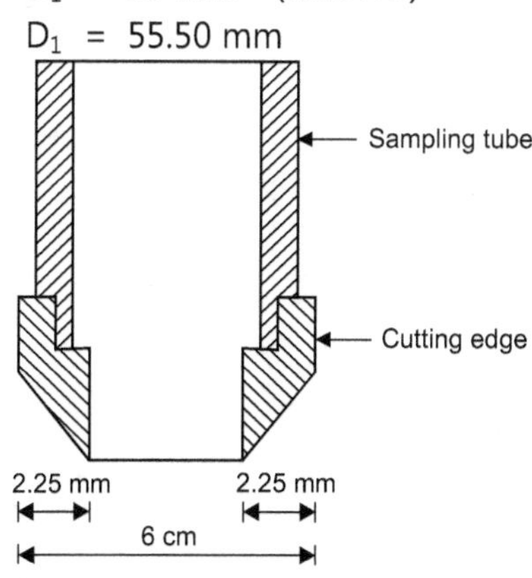

Fig. 1.21

(ii) Now, $A_r = \dfrac{(60)^2 - (55.50)^2}{(55.5)^2} \times 100$

A_r = 16.88%

(iii) As, area ratio is more than 10%, the sampler is not recommended for exploring undisturbed samples.

Example 1.2 : *Compute area ratio of a sampler with outer diameter 70 mm and thickness 2 mm. Comment.*

Solution :

(i)   Inner diameter (D_1)

= Outer diameter (D_2) – (Wall thickness × 2)

= 70 – (2 × 2) = 66 mm

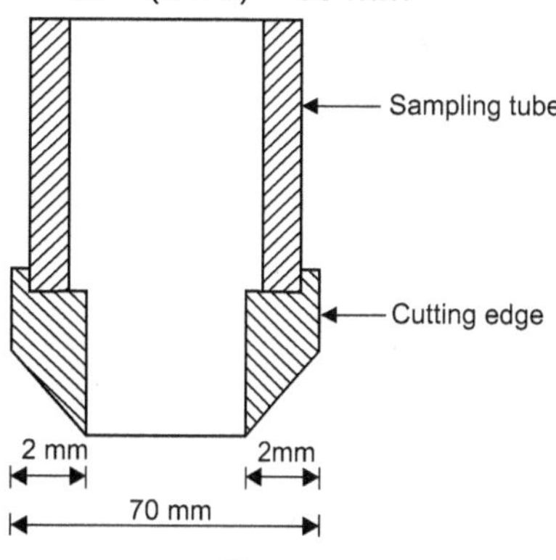

Fig. 1.22

(ii) Area ratio,

$$A_r = \frac{D_2^2 - D_1^2}{D_2^2} \times 100 = \frac{(70)^2 - (66)^2}{(66)^2} \times 100$$

A_r = 12.48 %

(iii) As, area ratio is more than 10%, sampler is not recommended for exploring undisturbed samples.

Example 1.3 : *A sampling tube of 15 cm internal diameter is 1 mm thick. It is fitted with a cutting. The inside diameter of cutting edge is flushed with sample tube. The cutting edge is 1.22 mm thick. Compute the inside and outside clearance and area ratio.*

Solution :

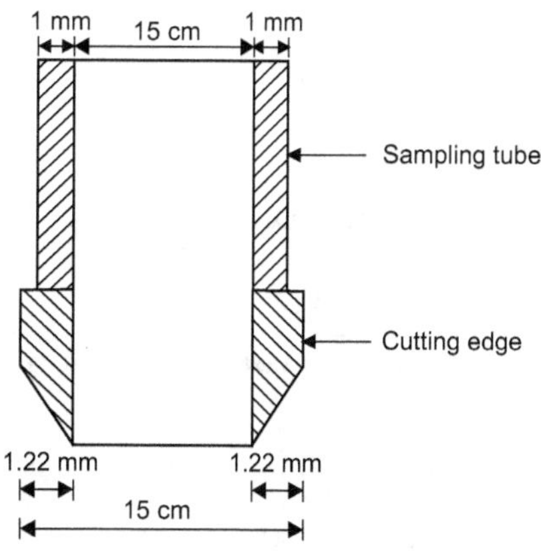

Fig. 1.23

In the above figure,

$$D_1 = 15 \text{ cm}$$
$$= \text{Inner diameter of cutting edge}$$
$$D_2 = \text{Outer diameter of cutting edge}$$
$$= 150 \text{ mm} + (2 \times 1.22)$$
$$D_2 = 152.44 \text{ mm}$$
$$D_3 = \text{Inner diameter of sampling tube}$$
$$= 15 \text{ cm} = 150 \text{ mm}$$
$$D_4 = \text{Outer diameter of sampling tube}$$
$$= 150 \text{ mm} + (2 \times 1 \text{ mm})$$
$$D_4 = 152 \text{ mm}$$

(i) Inside clearance

$$C_i = \frac{D_3 - D_1}{D_1} \times 100 = \frac{150 - 150}{150} \times 100$$
$$C_i = 0\%$$

(ii) Outside clearance,

$$C_o = \frac{D_2 - D_4}{D_4} \times 100 = \frac{152.44 - 152}{152} \times 100$$
$$C_o = 0.289\%$$

(iii) Area ratio,

$$A_r = \frac{D_2^2 - D_1^2}{D_1^2} \times 100 = \frac{(152.44)^2 - (150)^2}{(150)^2} \times 100$$
$$A_r = 3.27\%$$

Example 1.4: *A large number of undisturbed samples are to be obtained using 100 mm diameter (inside) sampling tube. Making suitable assumptions determine the thickness of sampling tube.*

Solution : (i) Assuming area ratio = 20%.

$$A_r = \frac{D_2^2 - D_1^2}{D_2^2} \times 100$$

$$20 = \frac{D_2^2 - (100)^2}{(100)^2} \times 100$$

$$D_2 = 109.54 \text{ mm}$$

(ii) Wall thickness

$$= \frac{1}{2}\,[D_2 - D_1] = \frac{1}{2} \times 9.54$$
$$= 4.77 \text{ mm}$$

Example 1.5 : *Determine area ratios of samplers of the following description.*

(i) Split-spoon sampler, D_o = 50 mm and D_i = 36 mm.

(ii) Drive tube, D_o = 100 mm and D_i = 85 mm.

(iii) Shelby tube, D_o = 50 mm and D_i = 45 mm.

Solution : Area ratio is given by,

$$A_r = \frac{D_2^2 - D_1^2}{D_1^2} \times 100 \ \text{ or }\ \frac{D_o^2 - D_i^2}{D_o^2} \times 100$$

$$D_2 \text{ or } D_o = \text{Outer diameter}$$
$$D_1 \text{ or } D_i = \text{Inner diameter}$$

(i) Split-spoon sampler,

$$A_r = \frac{(50)^2 - (36)^2}{(36)^2} \times 100$$
$$A_r = 92.90\%$$

(ii) Drive tube,

$$A_r = \frac{(100)^2 - (85)^2}{(85)^2} \times 100$$
$$A_r = 38.40\%$$

(iii) Shelby tube,

$$A_r = \frac{(50)^2 - (45)^2}{(45)^2} \times 100$$
$$A_r = 23.45\%$$

Example 1.6 : *The inner diameters of sampling tube and that of cutting edge are 70 mm and 68 mm respectively. Their outer diameters are 72 mm and 74 mm respectively. Determine inside clearance, outside clearance and area ratio of the sampler.*

Solution :

For Fig. 1.24 we have,

$$D_1 = 68 \text{ mm}$$
$$D_2 = 74 \text{ mm}$$
$$D_3 = 70 \text{ mm}$$
$$D_4 = 72 \text{ mm}$$

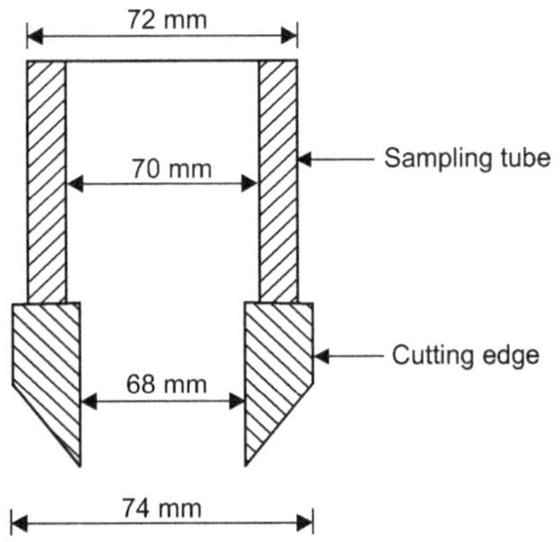

Fig. 1.24

(i) Inside clearance,

$$C_i = \frac{D_3 - D_1}{D_1} \times 100 = \frac{70 - 68}{68} \times 100$$
$$C_i = 2.941\%$$

(ii) Outside clearance,

$$C_o = \frac{D_2 - D_4}{D_4} \times 100 = \frac{74 - 72}{72} \times 100$$

$\therefore \quad C_o = 2.777\%$

(iii) Area ratio,

$$A_r = \frac{D_2^2 - D_1^2}{D_1^2} \times 100 = \frac{(74)^2 - (68)^2}{(68)^2} \times 100$$

$\therefore \quad A_r = 18.42\%$

Example 1.7 : *What is area ratio of 38 mm diameter, 2 mm thick sampling tube ?*

Solution : (i) As inner diameter is 38 mm.

$\therefore \quad D_1 = 38$

$\therefore$ Outer diameter,

$$D_2 = D_1 + (2 \times \text{Wall thickness})$$

$$= 38 + (2 \times 2) = 42 \text{ mm}$$

(ii) Area ratio,

$$A_r = \frac{D_2^2 - D_1^2}{D_1^2} \times 100 = \frac{(42)^2 - (38)^2}{(38)^2} \times 100$$

$$A_r = 22.16\%$$

Example 1.8 : *A SPT gives the average blow count of 32 in a fine saturated sand. What is corrected value of blow count ? Why is the measured value different ?*

Solution : (i) Here, $N_O = 32$ = Observed value.

Using relation,

$$N = 15 + 0.5(N_O - 15)$$

$\therefore \quad N = 15 + \frac{1}{2}(32 - 15)$

$\therefore \quad N = 15 + \frac{1}{2}(17)$

$\therefore \quad N = 23.5 = 24$

(ii) In case of fine saturated sand below the water table, developes pore pressure which is not easily dissipated. This pore pressure increases the resistance of soil and hence the penetration number (N').

Therefore, N and N' values are different.

Example 1.9 : *SPT is conducted in fine sand below water table and a value of 25 is obtained for N. What is correct value for 'N' ?*

Solution : (i) Here, $N_O = 25$ = Observed value of 'N'

Using relation,

$$N = 15 + 0.5 (N' - 15)$$

$\therefore \quad N = 15 + \frac{1}{2}(25 - 15)$

$$N = 15 + \frac{1}{2}(10)$$

$\therefore \quad N = 20$

Corrected value of N = 20.

Example 1.10: *A SPT was conducted in dense sand deposit at a depth of 22 m and value of 48 was obtained for 'N'. The density of sand was 15 kN/m³. What is the value of 'N'. Corrected for overburden pressure. ?*

Solution : (i)   s = Overburden pressure

$\therefore \quad$ s = Depth $\times$ Density of soil

$$= 22 \text{ m} \times 15 \text{ kN/m}^3$$

$$s = 330 \text{ kN/m}^2$$

(ii) Using relation,

$$N = N' \times \frac{350}{(s + 70)}$$

$$N = 48 \times \frac{350}{(330 + 70)}$$

$\therefore \quad N = 42$

Example 1.11: *What is corrected blow count in a fine saturated sand if the recorded blow count is 35 ?*

Solution : Here, N' = 35 = Observed blow count.

Using relation,

$$N = 15 + 0.5 (N' - 15)$$

$\therefore \quad N = 15 + \frac{1}{2}(35 - 15)$

$$N = 25$$

Example 1.12: *The field N value in a deposit of fully submerged fine sand was 36 at a depth of 6.5 m. The average saturated unit weight of the soil is 18.8 kN/m³. Calculate the corrected N value.*

Solution: (i) Overburden pressure,

$$s = \text{Depth} \times \text{Submerged density of soil}$$

$$= 6.5 \times [18.8 \text{ kN/m}^3 - 10 \text{ kN/m}^3]$$

$$= 6.5 \times (8.8)$$

$$= 57.2 \text{ kN/m}^2$$

(ii) N' = Observed value = 36.

Now, $N = N' \times \dfrac{350}{(s + 70)} = 36 \times \left(\dfrac{350}{57.2 + 70}\right)$

$\therefore \quad N = 99.05$

Example 1.13 : A seismic refraction study of an area has given the following data :

Distance from Impact Point to Geophones (m)	15	30	60	90	120
Time to Receive Wave (sec)	0.025	0.05	0.10	0.11	0.12

(i) Plot the time travel data and determine the seismic velocity for the surface layer and underlying layer.

(ii) Determine the thickness of the upper layer.

(iii) Using seismic velocity information, give the probable earth materials in the two layers.

Solution : (i) Figure 1.25 shows time-travel graph.

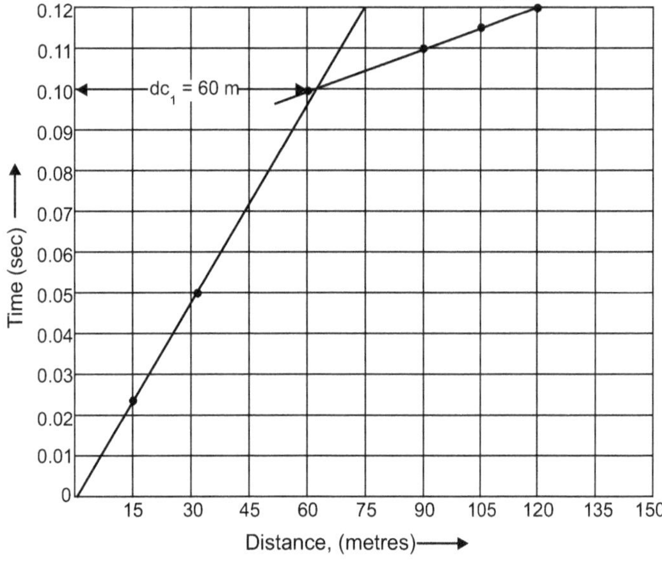

Fig. 1.25

From graph,

Critical distance, dc_1 = 60 m.

and Velocity in the upper layer , $v_1 = \left[\dfrac{(60 - 15)}{(0.10 - 0.025)}\right]$ = 600 m/sec

and Velocity in the lower layer , $v_2 = \left[\dfrac{(120 - 60)}{(0.12 - 0.10)}\right]$ = 3000 m/sec

(ii) Thickness of upper layer,

$$H_1 = \frac{dc}{2} \sqrt{\frac{v_2 - v_1}{v_2 + v_1}}$$

$\therefore \quad H_1 = \dfrac{60}{2} \times \sqrt{\dfrac{(3000 - 600)}{(3000 + 600)}}$

$\therefore \quad H_1 = 24.5$ m

(iii) From seismic velocity values, the probable materials are hard clay overlaying sound rock.

Example 1.14 : Calculate the depth of soft strata underlain by hard strata. Use the following data :

(i) Velocity in upper layer is half of velocity in lower layer.

(ii) Intersecting distance of velocity plots from time axis is 100 m.

Solution : (i) Velocity in upper layer = v_1,

Velocity in lower layer = v_2.

$$v_1 = \frac{1}{2} v_2 \qquad \text{... (given condition)}$$

And critical distance, dc = 100 m   (given)

(ii) Using relation,

$$H_1 = \frac{dc}{2} \times \sqrt{\frac{v_2 - v_1}{v_2 + v_1}}$$

as $v_1 = \dfrac{1}{2} v_2$

$\therefore \quad v_2 = 2v_1$

Putting value of v_2 in equation for H_1

$$H_1 = \frac{100}{2} \sqrt{\frac{(2v_1 - v_1)}{(2v_1 + v_1)}} = \frac{100}{2} \sqrt{\frac{v_1}{3v_1}}$$

$$= \frac{100}{2} \sqrt{\frac{1}{3}}$$

Depth of soft strata,

$H_1 = 28.86$ m

Example 1.15 : In a seismic exploration, the time distance plot gives velocity : v_1 = 580 m/s and v_2 = 4080 m/sec and break in the plot was noted at 30 m. Determine the depth of overburden above the underlying stiff layer.

Solution : Using relation,

$$H_1 = \frac{dc}{2} \sqrt{\frac{v_2 - v_1}{v_1 + v_2}}$$

$$= \frac{30}{2} \sqrt{\frac{4080 - 580}{4080 + 580}} = 12.999$$

$\therefore$ Depth of overburden,

$H_1 = 13$ m

Example 1.16 : The inner diameters of sampling tube and that of cutting edge are 72 mm and 70 mm respectively. Their outer diameters are 74 mm and 76 mm respectively. Determine the inside clearance, outside clearance and area ratio of sampler.

Solution : Sampling tube = D_i = 72 mm, cutting edge ϕ_i = 70 mm

D_0 = 74 mm, outer diameter edge = 76 mm

Area of cutting edge

$$= \frac{D_0^2 - D_i^2}{D_0^2} \times 100$$

$$= \frac{76^2 - 70^2}{76^2} \times 100 = 15.16 \%$$

We get,

$$\text{Inside clearance } (C_i) = \frac{D_0 - D_1}{D_i} \times 100$$

$$= \frac{74 - 72}{72} \times 100 = 2.78\%$$

Now find outside clearance,

Outside clearance (C_o)

$$= \frac{D_2 - D_4}{D_4} \times 100$$

$$= \frac{76 - 74}{74} \times 100 = 2.70\%$$

Area ratio of sampler

$$= \frac{D_0^2 - D_1^2}{D_0^2} \times 100$$

$$= \frac{74^2 - 72^2}{74^2} \times 100 = 5.33\%$$

EXERCISE

1. What are the objectives of soil investigation ?
2. State and explain the factors on which the extent of exploration of soil depends.
3. How will you plan soil exploration for an important building project ?
4. A multi-storeyed building is to be constructed on a bank of a river, when rock bed is expected at about 9 m depth. What will be your plan of action for geotechnical investigation ?
5. How will you determine the number and depth of exploratory holes for an important project.
6. What is significant depth ? How would you decide the depth of exploration ?
7. State and explain the factors on which the extent of exploration of soil depends.
8. What are the factors that influence the depth and the number of exploratory holes ?
9. What are the factors that influence the depth, number and lateral extent of site investigation ?
10. Explain the method 'Wash Boring' for taking samples and its reliability.
11. Describe various methods of drilling holes for sub-surface investigations.
12. Write short notes : Percussion drilling.
13. Describe with neat sketch, the method of wash boring used for soil exploration.
14. Explain in brief :
    (i) Auger boring.
    (ii) Wash boring.
    (iii) Percussion boring.

15. Describe various types of soil samples. What is area ratio ? State its significance.
16. Explain with sketches the terms
    (i) Inside clearance.
    (ii) Outside clearance.
    (iii) Area ratio.
    Comment upon its usefulness.
17. Distinguish between inside clearance and outside clearance. Draw sketches.
18. What is significance of area ratio applied to soil samplers ?
19. Explain the term area ratio with reference to sampling of soil alongwith their values and interpretation thereof.
20. What are the factors that effect the sample disturbance ? How are these effects minimised ?
21. What are the design features of soil sampler ?
22. Write short note on Spilt-spoon sampler.
23. Explain the procedure of undistized sampling using a piston sampler. Draw a neat sketch. Will this sampler procedure intact samples ideally, in sandy soils ?
24. Explain the working of piston sampler.
25. Make a neat dimensioned sketch of split spoon sampler. What is its use ?
26. Describe with a figure the piston sampler giving the details of components, procedure and use.
27. Explain the standard penetration test.
28. Explain the standard penetration test. What are the various corrections ?
29. Write short notes :
    (a) Standard penetration test.
    (b) Corrections to observed 'N' value.
    (c) Static cone test.
    (d) DCT.
30. Draw a neat sketch of standard penetration test apparatus. Show thereon components and indicate their function. What observations would you collect and what corrections are applied for their values ?
31. Point out similarities between cone penetration and SPT in tabular form only.
32. Draw a neat sketch of a cone penetration apparatus. Show thereon 10 components and indicate their function. What observations would you collect and what corrections are needed for their values ?
33. Draw a labelled sketch to illustrate SPT test.

34. Differentiate between :
    (a) SPT and DCPT.
    (b) SPT and SCPT.
35. Point out similarities and differences between cone penetration and standard penetration test in tabular form only.
36. Write with a diagram, the detail about SPT on the following points :
    (a) Procedure in 10 steps.
    (b) Correction in the observations.
    (c) Correlation of results.
37. How static cone penetration test differs from SPT ?
38. Draw a neat sketch of SCPT and explain how this test is performed in the field.
39. Write short note on Pressuremeter test.
40. What makes pressuremeter testing quite distinctive as compared to other field tests?
41. Explain seismic method or resistivity method of soil exploration.
42. Explain the seismic refraction method of exploration. Where do you recommend this method ?
43. Write short notes :
    (a) Electrical resistivity method.
    (b) Seismic refraction method.
44. Discuss geophysical methods for soil exploration.
45. If $D_1 T_1$ and $D_2 T_2$ are geophone distances and pick time observations respectively in seismic refraction method, explain with graphical presentation how would you obtain depths Z_1 and Z_2 of two layers. Also explain principle of this method.
46. Discuss seismic method in brief.
47. What are the different methods of soil investigation. Explain geo-physical method in detail.
48. Explain with sketches the layout and plot of seismic refraction method.
49. Explain with sketches seismic method of geophysical investigations giving details of principle, calculations and plotting of results.
50. Describe the procedure of resistivity method of soil exploration.

PROBLEMS FOR PRACTICE

1. Determine the area ratio of –
    (a) A tube sampler with, D_o = 165 mm and D_i = 150 mm.
    (b) A split-spoon with, D_o = 50.08 mm and D_i = 35 mm.
    (**Ans.** (a) 21%, (b) 104.7%)
2. A sampling tube has the outer diameter 78 mm and wall thickness 1.7 mm. Find area ratio. Comment on its suitability.
    (**Ans.** 9.4% < 10 suitable)
3. At depth of 4 m, blow count was 18. If the ground is saturated, find corrected value of N if unit weight of sound is 1.9 gm/cm^3.
    (**Ans.** N' = 32)
4. The blow count observed in a SPT conducted below water table in silty clay has 30. Find corrected blow count. At what depth overburden correction will be required if unit weight of soil is 14 kN/m^3.
    (**Ans.** N' = 22 count, at depth 5.96 m)
5. The blow count recorded at a depth of 12.5 m was 23. Correct the value for overburden pressure, if soil is moist sand. (γ = 18 kN/m^3)
    (**Ans.** N' = 27.28 ≈ 27.3)
6. In a geophysical exploration, the time distance plot gave, v_1 = 300 m/s and v_2 = 900 m/s and the break in the plot was located at 35 m. Determine the depth of overburden.
    (**Ans.** 16.93 m)
7. A SPT was performed at a depth of 20 m in a dense sand deposit with a unit weight of 17.5 kN/m^3. If observed N-value is 48. What is the corrected value for N ?
    (**Ans.** 40)
8. The inner diameters of a sampling tube and that of cutting edge are 70 mm and 68 mm respectively. Determine the inside clearance, outside clearance and area ratio of the sampler.
    (**Ans.** C_i = 2.94%, C_o = 2.78% and A_r = 18.43%)

◈ ◈ ◈

BEARING CAPACITY AND SETTLEMENT

2.1 INTRODUCTION

Foundation is the lowermost part of structure which is in direct contact with the ground and transmits the load of structure to the subsoil. The subsoil should have sufficient strength to bear the load being transmitted by foundation without rupture or shear failure. This capacity of soil is called its 'bearing capacity'.

2.1.1 Basic Terms

- **Bearing Capacity :** Capacity of the subsoil to bear a load.
- **Ultimate Bearing Capacity (q_u) :** It is the gross pressure at which base of foundation at soil fails in shear.
- **Net Ultimate Bearing Capacity (q_{nu}) :** It is the minimum net pressure (Gross pressure – Pressure due to self-weight of soil) at which soil fails in shear.

$$q_{nu} = q_u - \gamma d$$

- **Net Safe Bearing Capacity (q_{ns}):**

$$q_{ns} = \frac{q_{nu}}{F}$$

  where,   F : factor of safety (= 3)

- **Gross Safe Bearing Capacity or Safe Bearing Capacity (q_s) :** Maximum pressure which the soil can carry safely without risk of shear failure.

$$q_s = q_{ns} + \gamma d$$

  where, γ : unit weight of soil

  d : depth of footing

- **Net Safe Settlement Pressure (q_p) :** Maximum net pressure intensity of the loading, the soil can carry without exceeding allowable settlement.
- **Allowable Soil Pressure, ASP or Allowable Bearing Pressure (q_a) :** It is the net intensity of loading (maximum) at which the soil neither fails in shear nor there is excessive settlement.

2.2 MODES OF SHEAR FAILURE

Bearing capacity failures of foundations can be grouped into three categories as indicated below:

1. General shear failure
2. Local shear failure
3. Punching shear failure.

1. **General Shear Failure:**

This kind of failure occurs in soil possessing, brittle type stress-strain behaviour and is characterized by :

- Well defined failure pattern up to the ground surface.
- Sudden and catastrophic failure accompanied by tilting of base.
- Bulging of ground surface adjacent to foundation.
- Load settlement curve shows pronounced peak.
- Failure occur at very small vertical strain accompanied by large lateral strain.
- For granular soil - This failure is observed if soil is dense.
- For cohesive soil - This failure is observed if soil has stiff consistency.

2. **Local Shear Failure:**

This kind of failure occurs in soils possessing elasto-plastic stress-strain behaviour and is characterized by

- Well defined Wedge and slip surfaces only beneath the foundation.
- There is neither visible collapse nor substantial tilt of foundation.
- Slight bulging of ground surface adjacent to foundation.
- It is difficult to locate failure point (load) on load settlement curve.
- Failure occur at a very small lateral strain and large vertical strain.
- **For Granular Soil :** This failure is observed if soil has medium density.
- **For Cohesive Coil :** This failure is observed if soil has medium consistency.

3. **Punching Shear Failure:**

This kind of failure occurs in soil possessing plastic type stress-strain behaviour and is characterized by:

- Poorly defined slip surfaces.
- Footing sink into the soil.
- Soil beyond the loaded area is little or not affected. So there is no bulging of soil.
- Very difficult to locate the failure point on load settlement curve.

- Failure occurs at negligible or zero lateral strain and substantial vertical strain.
- **For Granular Soil :** Loose to very loose soil.
- **For Cohesive Soil :** Very soft to soft consistency.

Table 2.1 : Comparison of General Shear Failure and Local Shear Failure

Sr. No.	General Shear Failure	Local Shear Failure
1.	Occurs in dense/stiff soil.	Occurs in loose/soft soil.
2.	Soil parameters • Angle of internal friction $\varphi > 36°$ • Blow count N > 30 • Density index ID > 70% • Undrained cohesion Cu > 100kPa • Void ratio of soil e < 0.55	Soil parameters • Angle of internal friction $\varphi < 28°$ • Blow count N < 5 • Density index ID > 20% • Undrained cohesion Cu < 50kPa • Void ratio of soil e > 0.75
3.	Results in small vertical strain (<5%).	Results in large vertical strain (>20%).
4.	Failure pattern well defined & clear.	Failure pattern not well defined.
5.	Well defined peak in P-Δ curve.	No peak in P-Δ curve.
6.	Bulging formed in the neighbor-hood of footing at the surface.	Slight bulging observed in the neighbor-hood of footing.
7.	Extent of horizontal spread of disturbance at the surface large.	Extent of horizontal spread of disturbance at the surface very small.
8.	Observed in shallow foundations.	Observed in deep foundations.
9.	Failure is sudden & catastrophic.	Failure is gradual (progressive).
10.	Less settlement, but tilting failure observed.	Considerable settlement of footing observed.

Table 2.2 : Types of Failures and their Load Curves

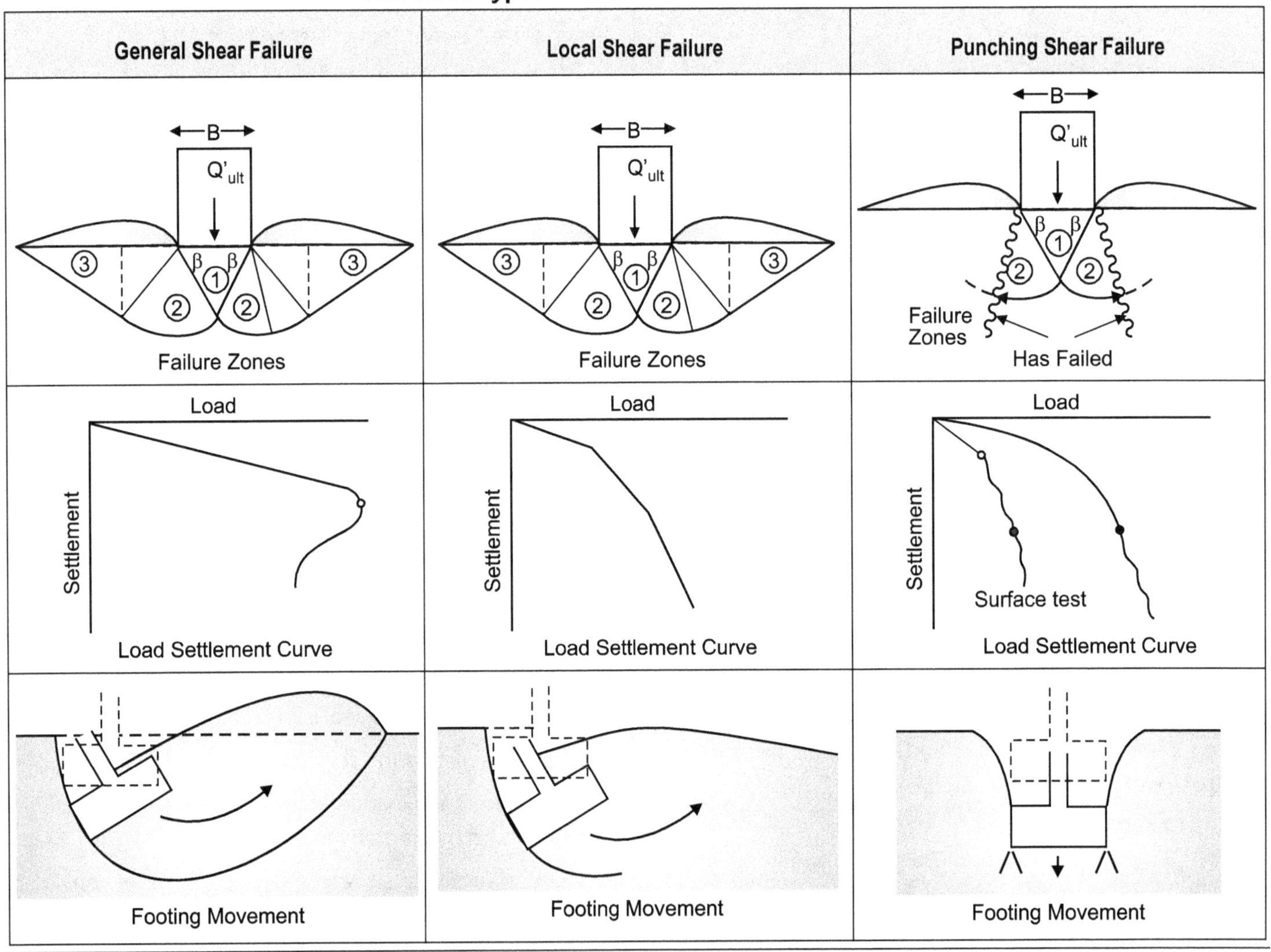

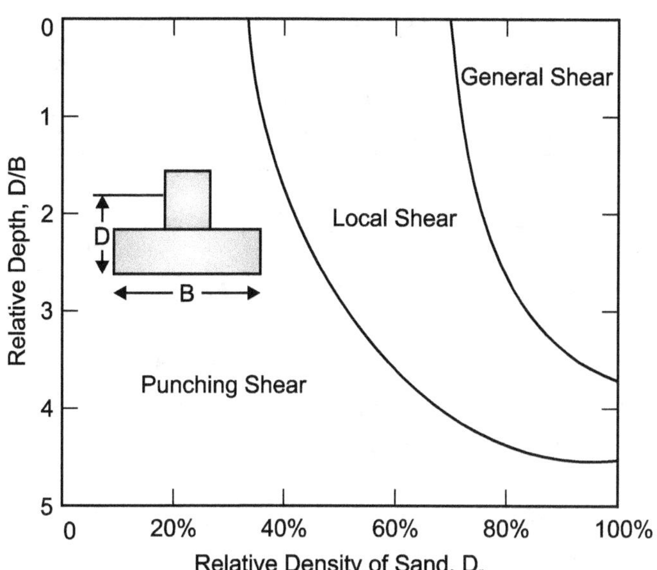

Fig. 2.1 : Types of Failure as a Function of Relative Density and Foundation Depth. (Vesic, 1963).

2.3 TERZAGHI'S BEARING CAPACITY EQUATION

Terzaghi gave a general theory of bearing capacity which is based on following assumptions :

- Soil is homogeneous and isotropic.
- Footing is laid at shallow depth.
- The shear strength of soil is represented by Mohr Coulombs Criteria.
- The footing is of strip footing type with rough base. It is essentially a two dimensional plane strain problem.
- Elastic zone has straight boundaries inclined at an angle equal to φ to the horizontal.
- Rupture surface extend only upto the level of base of footing.
- Shear resistance of soil above the base of footing is neglected.
- Method of superposition is valid.
- Passive pressure force has three components (PPC produced by cohesion, PPq produced by surcharge and PPγ produced by weight of shear zone).
- Effect of water table is neglected.
- Base of footing is rough.
- Footing carries axial and vertical loads.
- Footing and ground are horizontal.
- Limit equilibrium is reached simultaneously at all points.
- Complete shear failure is mobilized at all points at the same time.

- The properties of foundation soil do not change during the shear failure.

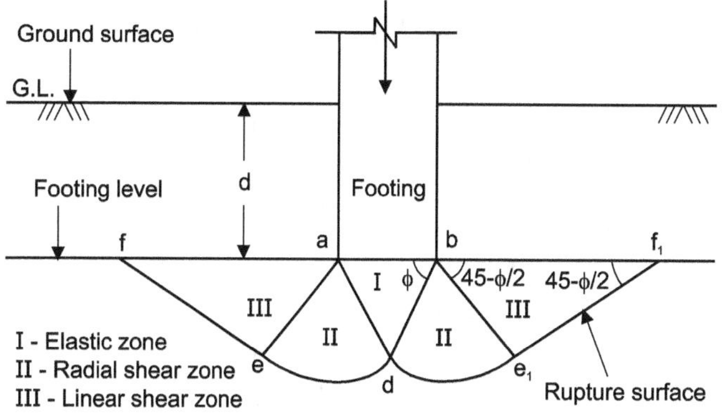

Fig. 2.2 : Terzaghi's analysis – Rupture surface

A strip footing of width B gradually compresses the foundation soil underneath due to the vertical load from superstructure. Let qf be the final load at which the foundation soil experiences failure due to the mobilization of plastic equilibrium. The foundation soil fails along the composite failure surface and the region is divided in to five zones :

- Zone 1 which is elastic,
- Two numbers of Zone 2 which are the zones of radial shear
- Two zones of Zone 3 which are the zones of linear shear.

Considering horizontal force equilibrium and incorporating empirical relation, the equation for ultimate bearing capacity is obtained as follows.

Let   F = Factor of safety (usually 3),

     c = cohesion, γ = unit weight of soil and φ – angle of internal friction

     D = Depth of foundation

     q = Surcharge at the base of footing

     B = Width of foundation

     Nc, Nq, Nγ = Bearing Capacity factors

Fig. 2.3 : Free body diagram of triangular wedge

This triangular wedge is subjected to following forces :

1. Contact pressure qu acting downward.

2. Self weight of wedge (W) acting downward.

3. Cohesion acting on inclined surface of wedge.

4. Passive pressure acting vertically on two inclined surface.

Applying equilibrium equation to the wedge i.e. (sum of forces in vertical direction is zero)

$$q_u(B \times 1) + W - 2C\sin\varphi - 2P_p = 0$$

$$W = \gamma\left[\frac{1}{2} B \frac{B}{2} \tan\varphi\right] = \frac{1}{4}\gamma B^2\tan\varphi$$

where,

C = Total cohesive force acting on each slant face of the wedge

$$= c \times \left[\frac{\beta}{2\cos\varphi}\right] \times 1$$

$$q_u(B \times 1) + \frac{1}{4}\gamma B^2\tan\varphi - 2\sin\varphi\left(C\frac{B}{2\cos\varphi}\right) - 2(P_{Pc} + P_{Pq}) = 0$$

Separating the terms containing Cohesion term, surcharge term and weight term i.e.

$$q_u B = [2P_{Pc} + Bc\tan\varphi] + [2P_{Pq}] + \left[2P_{Pr} - \frac{1}{4}\gamma B^2\tan\varphi\right]$$

$$q_u B = BcN_c + BqN_q + B(0.5r\,BN_\gamma)$$

$$q_u = cN_c + qN_q + (0.5r\,BN_\gamma)$$

This is the equation for calculating ultimate bearing capacity of strip footing

Values of Terzaghi's bearing capacity factors for various values of φ are as given below

Table 2.3 : Bearing Capacity Factor

φ	0	5	10	15	20	25	30	34	35	40	45	48	50
N_c	5.7	7.3	9.6	12.9	17.7	25.1	37.2	52.6	57.8	95.7	172.3	258.3	347.5
N_q	1.0	1.6	2.7	4.4	7.4	12.7	22.5	36.5	41.4	81.3	173.3	287.9	415.1
N_γ	0.0	0.5	1.2	2.5	5.0	9.7	19.7	36.0	42.4	100.4	297.5	780.1	1153.2

2.4 SPECIALIZATION OF TERZAGHI'S EQUATIONS

Terzaghi has derived an equation which is applicable for strip footing and for general shear failure of soil, later on this equation is modified to make it applicable for general case.

1. Modification of Equation to Account for Shape:

- To account for the shape of footing each term of the equation is multiplied by the shape factor and modified equation will then be

$$q_u = cN_cS_c + qN_qS_q + 0.5r\,BN_\gamma S_\gamma$$

- The values of these shape factors are as given below

Table 2.4

	Strip	Circular	Square	Rectangular
S_c	1	1.3	1.3	1 + 0.3(B/L)
S_q	1	1.0	1.0	1
S_γ	1	0.6	0.8	1 − 0.2(B/L)

2. Modification of Equation to Account for Local Shear Failure:

- To account for local shear type failure the equation is modified by modifying the value of shear parameters i.e. Replace c by c' φ by φ' and bearing capacity factors N by N' in the general equation given by Terzaghi where c' = 0.67 c and tan φ' = 0.67 tan φ. The values of bearing capacity factors for modified values of φ need to be calculated

- The values of bearing capacity factors for local shear failure are as given below

Table 2.5

φ	0	5	10	15	20	25	30	35	40	45	50
N_c'	5.7	6.7	8.0	9.7	11.8	14.8	19.0	25.2	34.9	51.2	81.3
N_q'	1.0	1.4	1.9	2.7	3.9	5.6	8.3	12.6	20.5	34.1	65.6
N_γ'	0.0	0.2	0.5	0.9	1.7	3.2	5.7	10.1	18.8	37.7	87.1

2.5 SKEMPTON'S FORMULA

Skempton proposed formula for purely cohesive soil

$$q_{nu} = cN_c$$

where, 　　c : cohesion

and 　　N_c : Skempton's bearing capacity factor

Case 1 : For Surface Footing (D = 0)

$$N_c = 5.14 - \text{for strip footing}$$

$$N_c = 6.20 - \text{for square or circular footing}$$

$$N_c = 5.14\left[1 + 0.2\frac{B}{L}\right] - \text{for rectangular footing}$$

Case 2 : For Footing with [D/B < 2.5]

$$N_c = \left[1 + 0.2\frac{D}{B}\right] \times [N_c]_{surface}$$

Case 3 : For Footing with [D/B > 2.5]

$$N_c = 1.5 \times [N_c]_{surface}$$

2.6 IS CODE METHOD (IS 6403-1981) FOR BEARING CAPACITY

- IS code has given an equation to calculate the bearing capacity Net ultimate bearing capacity is given by

$$q_{nu} = cN_cS_cd_ci_c + q(N_q - 1)S_qd_qi_q + 0.5\gamma B\,N_\gamma S_\gamma d_\gamma i_\gamma W'$$

... For general shear failure

$$q_{nu} = c'N_c'S_cd_ci_c + q(N_q' - 1)S_qd_qi_q + 0.5\gamma B\ N_\gamma'S_\gamma d_\gamma i_\gamma W'$$

$$\text{... For local shear failure}$$

where,

N_c, N_q, N_γ : Bearing capacity factors for general shear

N_c', N_q', N_γ' : Bearing capacity factors for local shear

Bearing capacity factors for general shear failure can be obtained from the table given below

Table 2.6

φ	0	5	10	15	20	25	30	35	40	45	50
N_c	5.14	6.49	8.35	10.98	14.38	20.72	30.14	46.12	75.31	138.88	266.89
N_q	1.0	1.57	2.47	3.94	6.40	10.66	18.40	33.30	64.20	134.88	319.07
N_γ	0.00	0.45	1.22	2.65	5.39	10.88	22.40	48.03	109.41	217.76	762.89

Note : For obtaining values of N_c', N_q', N_γ' can be calculated by calculating the value of φ' use this value to calculate the values of bearing capacity factors from above table instead of φ.

The values of **shape factor** for various shapes of footing is as given below

Table 2.7 : Shape Factor

	Strip	Circular	Square	Rectangular
S_c	1.0	1.3	1.3	1.0 + 0.2 (B/L)
S_q	1.0	1.2	1.2	1.0 + 0.2 (B/L)
S_γ	1.0	0.6	0.8	1.0 – 0.4 (B/L)

The values of **Depth Factor** can be calculated by following equation

$$d_c = 1 + 0.2\frac{D}{B}\sqrt{N_\varphi}$$

$$d_q = d_\gamma = 1 \text{ for } \varphi < 10°$$

$$d_q = d\gamma = 1 + 0.1\frac{D}{B}\sqrt{N_\varphi} \text{ for } \varphi > 10°$$

where, $N_\varphi = \tan^2\left(45 + \dfrac{\varphi}{2}\right)$

Note: Depth correction is to be applied only when backfilling is done with proper compaction.

The values of **inclination factor** can be calculated by following equation

$$i_c = i_q = \left(1 - \frac{\alpha}{90}\right)^2 \text{ and } i_\gamma = \left(1 - \frac{\alpha}{\varphi}\right)^2$$

where α : inclination of load with vertical.

2.7 MEYERHOFF'S METHOD

Meyerhoff equation for ultimate bearing capacity of footing is as below

$$q_u = cN_cS_cd_ci_c + qN_qS_qd_qi_q + 0.5\gamma\ BN_\gamma S_\gamma d_\gamma i_\gamma$$

where,　C : cohesion of soil (kPa),

B : width of footing (m),

D : depth of footing (m),

γ : Unit weight of soil (kN/m^3),

φ : Angle of internal friction for the soil

N : Meyerhoff's bearing capacity factors and are given by

$$N_q = e^{\pi\tan\varphi}\ K_p,$$

$$N_c = (N_q - 1)\cot\varphi$$

and　　$N_\gamma = (N_q - 1)\tan(1.4\varphi)$

Table 2.8 : Shape Depth and Inclination Factors

Factors	Value	Applicable when φ is
Shape	$S_c = 1 + 0.2\ K_p\dfrac{B}{L}$	Any value
	$S_q = S_\gamma = 1 + 0.1\ K_p\dfrac{B}{L}$	More than 10^0
	$S_q = S_\gamma = 1$	Equal to zero
Depth	$d_c = 1 + 0.2\sqrt{K_p}\dfrac{D}{B}$	Any value
	$d_q = d_\gamma = 1 + 0.1\sqrt{K_p}\dfrac{D}{B}$	More than 10^0
	$d_q = d_\gamma = 1$	Equal to zero
Inclination factors Inclination Factors	$i_c = i_q = \left(1 - \dfrac{\alpha}{90}\right)^2$	Any value
	$i_\gamma = \left(1 - \dfrac{\alpha}{\varphi}\right)^2$	More than 0^0
	$i_\gamma = 0$	Equal to zero

Table 2.9 : Meyerhoffs Bearing Capacity Factors

φ	0	5	10	15	20	25	30	35	40	45	50
N_c	5.14	6.5	8.3	11.0	14.8	20.7	30.1	46.1	75.3	133.9	266.9
N_q	1.0	1.6	2.5	3.9	6.4	10.7	18.4	33.3	64.2	134.9	319.0
N_γ	0	0.1	0.4	1.1	2.9	6.8	15.7	37.1	93.7	262.7	873.7

2.8 EFFECT OF POSITION OF WATER TABLE

- The effective unit weight of the soil is used in the bearing-capacity equations for computing the ultimate capacity. This has already been defined for **q** in the **qNq** term.

- Also wedge term $0.5\gamma BN_\gamma$ also uses the effective unit weight for the soil. Presence of water table in the vicinity of footing will decrease the bearing capacity of soil its effect is accounted either by introducing water table correction factors in the equation or by using equivalent unit weight of soil

Method 1 : Using Water Table Correction Factors :

- Multiply second term of the equation with water table correction factor [W_q or R_{w1}] and third term of the equation with water table correction factor [W_γ or R_{w2}].

i.e. by $\quad q_u = cN_cS_c + qN_qS_qW_q + 0.5\gamma BN_\gamma S_\gamma W_\gamma$

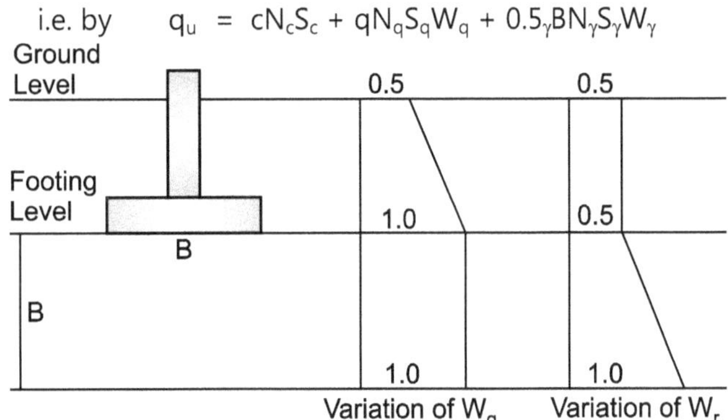

Fig. 2.4

Method 2 : Using Equivalent Unit Weight for the Soil:

- In this method weighted unit weight (equivalent unit weight) is determined. For soil above water table bulk unit weight is considered and for soil below water table submerged unit weight is considered.

When Water Table is in between Ground Level and Footing Level :

For surcharge term:

$$q = \gamma_e D = \gamma d + \gamma'(D - d)$$

$$\gamma_e = \frac{\gamma d + \gamma'(D - d)}{D} = \gamma' + \frac{d}{D}(\gamma - \gamma')$$

For weight term :

$$\gamma B = \gamma_e B = \gamma' B \text{ i.e. } \gamma_e = \gamma'$$

When water table is below footing

$$q = \gamma_e D = \gamma d \text{ i.e. } \gamma_e = \gamma$$

$$\gamma B = \gamma_e B = \gamma b + \gamma'(B - b)$$

i.e. $\quad \gamma_e = \dfrac{\gamma b + \gamma'(B - b)}{B} = \gamma + \dfrac{b}{B}(\gamma - \gamma')$

2.9 EFFECT OF ECCENTRICITY OF LOAD

When the load acting on footing is not axial we need to modify the dimensions of footing in such a way that for those modified dimensions the load will be axial. These reduced dimensions need to be used in place of original dimensions. The modified dimensions are given by original dimension – twice the eccentricity in that direction. These dimensions need to be used while calculating various factors of the bearing capacity equation

$$L' = L - 2e_L$$
$$B' = B - 2e_B$$

and $\quad A' = L' \times B'$

2.10 SETTLEMENT

- All structures that civil Engineers construct are founded on soil / rock. These structures apply load on the soil. When soil has to carry additional load, the soil particles have to adjust themselves and regroup themselves in a closer packing to withstand that load. This adjustment and regrouping results in causing the structures to settle.

- The vertical downward movement of a loaded base is called settlement.

- The external load acting on soil mass produces a time dependent accumulation of particle rolling, sliding, crushing and elastic distortion in a limited influence zone, beneath the loaded area. The statistical accumulation of movement in the direction of interest is called settlement.

2.10.1 Causes of Settlement

- The settlement (decrease in volume of a soil mass) is due to decrease in the volume of voids in the soil by expulsion of air, expulsion of water, compression of air, compression of soil solid and distortion of adsorbed water.

Table : 2.10

Internal Factor		External Factor
1. Compression of air		1. Static load
2. Expulsion of air	Initial	2. Dynamic load - (Machinery pile driving, explosion, earthquake etc.)
3. Rearrangement of particles(sliding, rolling)		3. W.T. fluctuation (stress variation and shrinkage)
4. Expulsion of water	Primary	4. Neighboring construction
5. Compression of soil particles		5. Other causes :
6. Reorientation of adsorbed water	Secondary	• Underground erosion • Thermal changes • Frost heave • Mining subsidence

2.10.2 Components of Settlement

- Total settlement of foundation is sum total of three components

    1. Initial settlement or elastic settlement,
    2. Primary consolidation settlement and
    3. Secondary consolidation settlement

Table 2.11 : Mechanism of Settlement of Structures

Mechanism	Nature	Duration
Elastic	Due to change in shape of stressed soil zone at constant volume.	Immediate
Compression of air	In partially saturated soil only, due to reduction of volume of air voids	Immediate
Consolidation Primary	Due to flow of water from voids of soil resulting in rearrangement of soil particles in a packing suitable to carry load imposed by the structure	Function of permeability and compressibility of soil
Secondary consolidation	Due to creep like behaviour of soil	Very slow

- In general settlement depends on following factors

    ➤ Magnitude of the stresses applied.
    ➤ Geometry of plan area which is stressed.
    ➤ Depth at which stress is applied in relation to width of stressed area (D/B), and the thickness of compressible stratum.
    ➤ Compressibility of stratum.
    ➤ Rigidity of the foundation element.

Table 2.12 : Types of Movement in Different Soils

Principal Soil Types	Type of Movement			
	Immediate	Consolidation	Creep	Swell
Rock	Yes	No	No	Some
Gravel	Yes	No	No	No
Sand	Yes	No	No	No
Silt	Yes	Minor	No	Yes
Clay	Yes	Yes	No	Yes
Organic	Yes	Minor	No	Yes

1. Elastic Settlement :

- The settlement which will take place immediately after applying load is called elastic settlement or immediate settlement.

- It occurs on account of elastic behaviour that produces distortion at constant volume.

- It is determined from elastic theory; it occurs in all types of soil due to elastic compression.

- It depends on the elastic properties of foundation soil, rigidity, size and shape of foundation.

It is computed by using

$$S_i = q\, B \left(\frac{1 - \mu^2}{E}\right) I$$

where,　　q : contact pressure at base of footing,

　　　　　B : width of footing,

　　　　　I : influence coefficient,

　　　　　E : modulus of elasticity of soil,

　　　　　μ : poisons ratio for the soil.

Table : 2.13

Typical Range of Poisson's Ratio	
Soil Type	**Poisson's Ratio**
Saturated clay	0.5
Sandy clay	0.3 – 0.4
Unsaturated clay	0.35 – 0.4
Loses	0.44
Silt	0.3 – 0.35
Sand	0.15 – 0.30
Rock	0.1 – 0.4

Table 2.14

Soil Modulus for Various Soil	
Soil Type	**Soil Modulus (kPa)**
Very soft clay	400 – 3000
Soft clay	1500 – 4000
Medium clay	3000 – 8500
Hard clay	7000 – 17000
Sandy clay	28000 – 42000

The values of Poisons ratio and Modulus of Elasticity values are only the representative values and can be used only in the absence of more accurate data. The influence factor I depends on the shape and flexibility of footing. Further, in flexible footing I is not constant. Table presents the different values of I for different soils.

Table 2.15 : Influence Factor (I)

Shape of Footing	Flexible			Rigid
	Center	Corner	Average	
Circle	1	0.64	0.85	0.8
Square	1.12	0.56	0.95	0.9
Rectangle L/B = 1.5	1.36	0.68	1.20	1.09
L/B = 2	1.52	0.77	1.31	1.22
L/B = 5	2.1	1.05	1.83	1.68
L/B = 10	2.52	1.26	2.25	2.02
L/B = 100	3.38	1.69	2.96	2.70

2. Consolidation Settlement:

- The settlement of a soil which takes place due to expulsion of water from the soil and is mainly applicable to saturated or nearly saturated clay.

- It occurs due to the process of consolidation, clay and organic soil are most prone to consolidation settlement.

- Consolidation is the process of reduction in volume due to expulsion of water under an increased load, it is a time related process occurring in saturated soil by draining water from void, it is often confused with compaction.

- Consolidation theory is required to predict both rate and magnitude of settlement.

- This settlement is given by

$$S = C_c \frac{H}{1 + e} \log_{10} \frac{P + \Delta P}{\Delta P}$$

Or

$$S = \frac{a_v}{1 + e} \Delta P \times H = m_v \times \Delta P \times H$$

where,

 S : Settlement,

 H : Thickness of compressible stratum,

 e : Initial void ratio,

 C_c : compression index,

 a_v : coefficient of compressibility,

 m_v : coefficient of volume compressibility,

 P : Initial pressure at center of layer (Geostatic stress),

 ΔP : Increase in pressure at centre of layer due to load on footing (stress due to external load)

3. Secondary Compression:

- The settlement which occurs at very slow and negligible rate after completion of initial consolidation is called **Secondary Consolidation or Creep Settlement.**

- This settlement starts after the primary consolidation is completely over.

- During this settlement, excess pore water pressure is zero.

- The reasons for secondary settlement are not clear.

- This is creep settlement occurring due to the readjustment of particles to a stable equilibrium under sustained loading over a long time.

- This settlement is common in very sensitive clay, organic soils and loose sand with clay binders.

It is calculated by using,

$$S_s = C_\alpha H \log \left(\frac{t_p + \Delta t}{t_p} \right)$$

where C_α : Coefficient of secondary compression

    H : Thickness of clay layer

    Δt : Time increment producing S_s

    t_p : Time taken for primary consolidation to complete

Table 2.16 : Typical Values of C_α

Type of Soil	C_α
Normally consolidated clays	0.005 – 0.02
Very plastic soils, organic soils	> = 0.03
Pre compressed clays with OCR > 2	< 0.001

2.10.3 Types of Settlement

1. Uniform Settlement: If every part of structure settles by same amount.

2. Non-Uniform Settlement:

- If different parts of the structure settle by unequal amount then it is called Non-uniform settlement and the difference between settlement at any two points is called **Differential settlement**.

- Differential settlement per unit length is called Tilt or Angular distortion.

Causes of Differential Settlement

- Large loaded areas on flexible foundation.

- Non-homogeneity in the soil.

- Non-uniform loading.

- Difference in time of construction of adjacent parts of structure.

- Overlapping of stresses from adjoining structures.

3. Tolerable Settlement or Allowable Settlement :

- The settlement of foundation which will not cause any effect on the performance of structure is called allowable settlement or permissible settlement of foundation.

- Magnitude of this settlement depends on the type of foundation (Isolated / rigid) and type of soil (clay / sand). These values will be specified by IS code

2.10.4 Effect of Settlement

As long as settlement is uniform, no damage will be done to structure. The only effect it can have is on the service lines, such as water and sanitary pipe, telephone and electric cables etc. which can snap (Break suddenly) if settlement is considerable. Building in Mexico City have undergone such settlement in a large measure where the first floor, it seems has come down to ground floor level.

If settlement is non-uniform it will produce harmful effects:

- Failure of pavement as a result of depression at corners and transverse joint.

- Adverse effects on gradient of railways.

- Cracking of plaster, displacement of door and window frames, loosening of brick (masonry).

- Unequal settlement of pier of bridge causing traffic hazards.

- In statically indeterminate structures (beams, frames, arches, vaults etc.). Settlement induces additional moments which will induce tension cracks in structure.

2.10.5 Methods of Reducing Settlement

- Total settlement or Uniform settlement
  - ➤ Preloading
  - ➤ By using floating foundation
- Differential settlement
  - ➤ By providing deep foundation
  - ➤ By providing rigid material foundation
  - ➤ Stabilization of soil
  - ➤ By reinforcing the soil
  - ➤ By providing additional loads on lightly loaded area
  - ➤ By eccentrically loaded mass
  - ➤ By increasing the rigidity of structure by providing column brackets.

SOLVED EXAMPLES

Example 2.1 : *Determine ultimate bearing capacity of strip footing of width 1.5 m resting at a depth of 1 m below ground on sandy soil under following water table conditions :*

   (i)   Water table at a depth of 0.5 m below ground level.

   (ii)   Water table at a depth of a 0.5 m below base of footing.

Assume γ_{sat} = 20 kN/m³, γ = 17 kN/m³, ϕ = 30°, N_q = 60, N_γ = 75.

Solution : Case (i) :

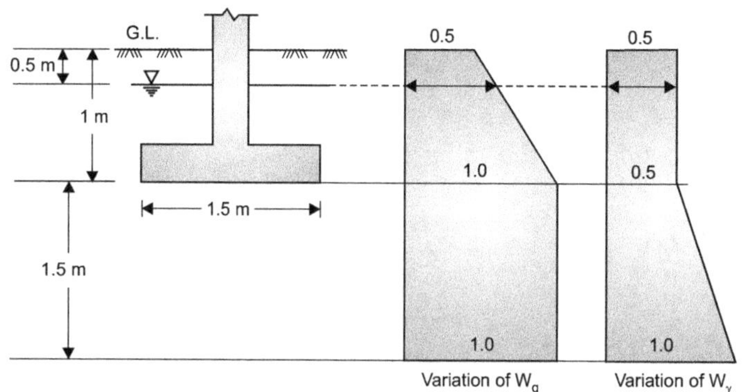

Fig. 2.5

$$W_q = 0.5 + \frac{0.5}{1.0}\,(0.5) = 0.75$$

$$W_\gamma = 0.5$$

Ultimate bearing capacity of strip footing,

$$q_u = cN_c + \gamma_1\, dN_qW_q + 0.5\,\gamma_2\, BN_\gamma W_\gamma$$

$$\gamma_1 = \frac{0.5\,(17) + 0.5\,(20)}{1.0}$$

$$= 18.5 \text{ kN/m}^3$$

$$W_q = 0.75 \text{ and}$$

$$W_\gamma = 0.5 \text{ (from diagram of } W_q \text{ and } W_\gamma)$$

$$\therefore \quad q_u = 0 + (18.5)\,(1.0)\,(60)\,(0.75)$$

$$+ (0.5)\,(20)\,(1.5)\,(75)\,(0.5)$$

$$= 1395 \text{ kN/m}^2$$

Case (ii) :

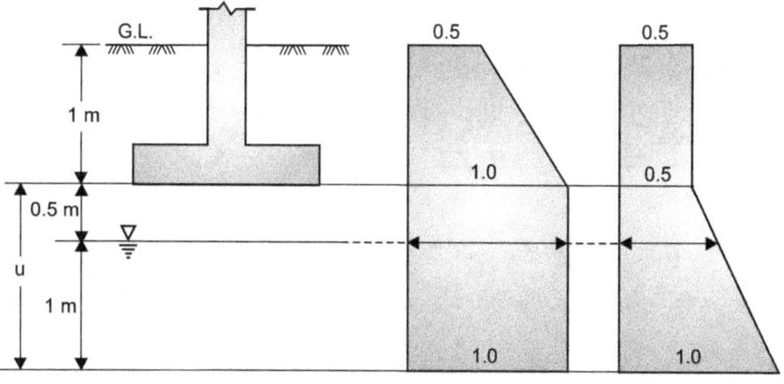

Fig. 2.6

$$W_q = 1.0$$

$$W_\gamma = 0.5 + \frac{0.5}{1.5}\,(0.5)$$

$$= 0.67$$

$$\gamma_1 = 17\ kN/m^3 \text{ (above footing)}$$

$$\gamma_2 = \frac{0.5\,(17) + 1.0\,(20)}{1.5}$$

$$= 19\ kN/m^3$$

$$q_u = 0 + (17)\,(1.0)\,(60)\,(1.0)$$

$$\quad + 0.5\,(19)\,(1.5)\,(75)\,(0.67)$$

$$= 1736.06\ kN/m^2$$

Example 2.2 : *A footing 2 m square rests on soft clay soil with its base at a depth of 1.5 m from ground surface. Using Skempton's equation determine net safe bearing capacity of the footing. For the soil properties are $C_u = 50\ kN/m^2$ and $\phi_u = 0$.*

Solution :

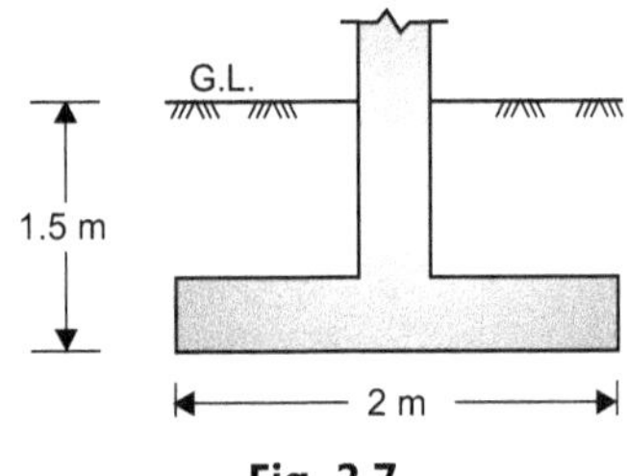

Fig. 2.7

Net ultimate bearing capacity of square footing (Skempton's equation)

$$q_{nu} = cN_c$$

$$\text{for} \quad d < 2.5\ B$$

$$N_c = \left(1 + 0.2\,\frac{d}{B}\right)\ (6.2)$$

$$= \left[1 + 0.2\left(\frac{1.5}{2.0}\right)\right]\ (6.2)$$

$$= 7.13$$

$$q_{nu} = (50)\,(7.13) = 356.5\ kN/m^2$$

Net safe bearing capacity (assuming factor of safety = 3)

$$q_{ns} = \frac{q_{nu}}{F} = \frac{356.5}{3.0} = 118.83\ kN/m^2$$

Example 2.3 : *A strip footing, 1 m wide at its base is located at a depth of 0.8 m below the ground surface. The properties of foundation soil are $\gamma = 18\ kN/m^3$, $c = 20\ kN/m^2$ and $\phi = 20°$.*

Determine safe bearing capacity, using F.S. = 3.0.

Use Terzaghi's analysis. Assume soil fails by local shear. For

$\phi = 20°$, $N_c' = 11.8$, $N_q' = 3.9$ and $N_\gamma' = 1.7$.

Solution :

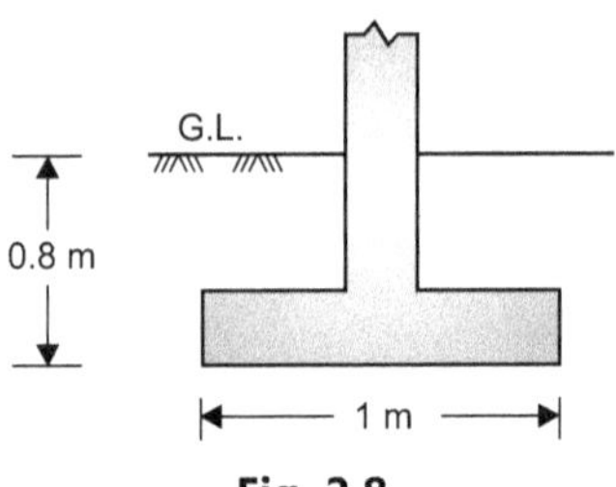

Fig. 2.8

Terzaghi's equation for local shear failure (strip footing) :

Ultimate bearing capacity,

$$q_u = c'N_c' + \gamma_1 dN_q' + 0.5\,\gamma_2\,BN_\gamma'$$

Here, $\quad c' = \frac{2}{3}\,(30) = 20\ kN/m^2$

$$\therefore \quad q_u = 20(11.8) + 18(0.8)\,(3.9) + 0.5(18)\,(1)\,(1.7)$$

$$= 307.46\ kN/m^2$$

Net ultimate bearing capacity

$$q_{nu} = q_u - \gamma_1 \cdot d$$

$$= 307.46 - 18(0.8)$$

$$= 293.06\ kN/m^2$$

Net safe bearing capacity

$$q_{ns} = \frac{q_{nu}}{F} = \frac{293.06}{3.0} = 97.68\ kN/m^2$$

Safe bearing capacity

$$q_s = q_{ns} + \gamma_1 \cdot d$$

$$= 97.68 + 18(0.8)$$

$$= 112.08\ kN/m^2$$

Example 2.4 : *A loading test was conducted with 300 mm square plate at depth of 1 m below the ground surface. In a pure clay deposed water table is located at a depth of 4 m below ground level. Failure occurred at a load of 45 kN. What is safe bearing capacity of a 1.5 m wide strip footing at 1.5 m depth in same soil ? Assume $\gamma = 18\ kN/m^3$ above water table and factor of safety of 2.5.*

Solution : As water table is at large depth (more than width of footing below footing) so has no effect on bearing capacity.

For clayey soil,

$$q_{uf} = q_{up} \qquad \text{(f – footing, p – plate)}$$

Ultimate bearing capacity of footing $= \dfrac{45}{(0.3)\,(0.3)} = 500\ kN/m^2$

Net ultimate bearing capacity,

$$q_{nu} = q_u - \gamma \cdot d$$

$$= 500 - 18(1.5) = 473\ kN/m^2$$

Net safe bearing capacity,

$$q_{ns} = \frac{q_{nu}}{F} = \frac{473}{2.5} = 189.2 \text{ kN/m}^2$$

Safe bearing capacity,

$$q_s = q_{ns} + \gamma \cdot d$$
$$= 189.2 + 18(1.5)$$
$$= 216.2 \text{ kN/m}^2$$

Example 2.5 : *Compute safe bearing capacity of a continuous footing 1.8 m wide and located at a depth of 1.2 m below ground level in a soil having unit weight γ = 20 kN/m^3, c = 20 kN/m^2 and ϕ = 20°. Assume factor of safety of 2.5. Terzaghi's bearing capacity factors for ϕ = 20° are N_c = 17.7, N_q = 7.4 and N_γ = 5.0, what is permissible load per metre run of the footing ?*

Solution : For continuous (strip) footing

$$q_u = cN_c + \gamma_1 dN_q + 0.5\,\gamma_2 BN_\gamma$$
$$= 20(17.7) + 20(1.2)(7.4) + 0.5(20)(1.8)(5.0)$$
$$= 621.6 \text{ kN/m}^2$$

$$q_{nu} = q_u - \gamma \cdot d$$
$$= 621.6 - 20(1.2)$$
$$= 597.6 \text{ kN/m}^2$$

$$q_{ns} = \frac{q_{nu}}{F} = \frac{597.6}{2.5} = 239.04 \text{ kN/m}^2$$

$$q_s = q_{ns} + \gamma \cdot d = 239.04 + 20(1.2)$$
$$= 263.04 \text{ kN/m}^2$$

Permissible load per metre run of footing,

$$= \text{(Safe bearing capacity) (Width of footing)}$$
$$= (263.04)(1.8)$$
$$= 473.47 \text{ kN/m}$$
$$\cong 470 \text{ kN/m}$$

Example 2.6 : *Compute safe bearing capacity of a square footing 1.5 m × 1.5 m located at a depth of 1 m below ground level in sandy soil of average density 20 kN/m^3, ϕ = 20°, N_c = 17.7, N_q = 7.4 and N_γ = 5.0. Assume suitable factor of safety and that water table is very deep. Also compute reduction in safe bearing capacity of footing if water table rises to the ground level.*

Solution : For square footing :

$$q_u = 1.3\,cN_c + \gamma_1 dN_q + 0.4\,\gamma_2 BN_\gamma$$
$$= (1.3)(0)(17.7) + 20(1)(7.4)$$
$$+ 0.4(20)(1.5)(5.0)$$
$$= 208 \text{ kN/m}^2$$

$$q_{nu} = q_u - \gamma \cdot d$$
$$= 208 - 20(1) = 188 \text{ kN/m}^2$$

$$q_{ns} = \frac{q_{nu}}{F} = \frac{188}{3.0}$$
$$= 62.67 \text{ kN/m}^2 \text{ (assuming F.S. = 3.0)}$$

$$q_s = q_{ns} + \gamma \cdot d$$
$$= 62.67 + 20(1.0)$$
$$= 82.67 \text{ kN/m}^2$$

When water table is at ground level

$$q_u = 1.3\,cN_c + \gamma_1 dN_q W_q + 0.4\,\gamma_2 BN_\gamma W_\gamma$$
$$= 0 + 20(1)(7.4)(0.5) + 0.4(20)(1.5)(5.0)(0.5)$$
$$= 104 \text{ kN/m}^2$$

$$q_{nu} = q_u - \gamma'd = 104 - 10(1) = 94 \text{ kN/m}^2$$

$$q_{ns} = \frac{q_{nu}}{F} = \frac{94.0}{3.0} = 31.33 \text{ kN/m}^2$$

$$q_s = q_{ns} + \gamma'd$$
$$= 31.33 + 10(1) = 41.33 \text{ kN/m}^2$$

Reduction in bearing capacity,

$$82.67 - 41.33 = 41.34 \text{ kN/m}^2 \quad \text{(50\% reduction)}$$

Example 2.7 : *Compute the ultimate load that an eccentrically loaded square footing of width 2.1 m with an eccentricity of 0.35 can carry at a depth of 0.5 m in a soil with γ = 18 kN/m^2, c = 9 kN/m^2 and ϕ = 36°, N_c = 52, N_q = 35 and N_γ = 42. (Adopt useful with concept).*

Solution : Effective width with useful width

$$B_e = B - 2e = 2.1 - 2(0.35) = 1.4 \text{ m}$$

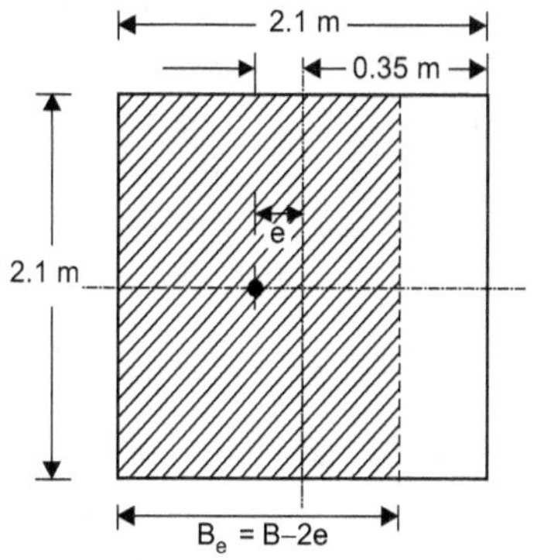

Fig. 2.9

For square footing, ultimate bearing capacity

$$q_u = 1.3\,cN_c + \gamma_1 d N_q + 0.4\,\gamma_2 B N_\gamma$$

For eccentrically loaded footing B is replaced by B_e.

$$q_u = 1.3(9)(52) + 18(0.5)(35)$$
$$+ 0.4(18)(1.4)(42)$$
$$= 965.74 \text{ kN/m}^2$$

Example 2.8 : *A plate load test was carried using plate of size 30 cm × 30 cm on cohesionless soil from the necessary graph drawn, it was noticed that for a settlement of 10 mm of plate intensity of load was 160 kN/m² if permissible settlement of foundation is 20 mm. Determine the load that can be applied on a footing of size 2 m × 2 m.*

Solution: Settlement of plate for permissible settlement of foundation Δ_f is

$$\Delta_p = \frac{4}{3}\,\Delta_f \left[\frac{B_p\,(B_f + 0.3)}{B_f\,(B_p + 0.3)}\right]^2$$

$$= \frac{4}{3}\,(20) \left[\frac{0.3\,(2 + 0.3)}{2\,(0.3 + 0.3)}\right]^2$$

$$= 8.82 \text{ mm } (< 10 \text{ mm for which load was}$$

scaled from graph)

$$q_f = q_p \frac{B_f}{B_p} = 160 \left(\frac{2.0}{0.3}\right)$$

$$= 1066.67 \text{ kN/m}^2$$

Ultimate bearing capacity of footing = 1066.67 kN/m²

$$q_{nu} = q_u - \gamma \cdot d$$

$$q_{ns} = \frac{q_{nu}}{F} = \frac{q_u - \gamma \cdot d}{F}$$

$$q_s = q_{ns} + \gamma \cdot d = \frac{q_u - \gamma d}{F} + \gamma \cdot d$$

$$\cong \frac{q_u}{F} = \frac{1066.67}{3}$$

$$= 355.55 \text{ kN/m}^2 = 350$$

Safe load on footing = (350) (2 × 2) = 1400 kN/m²

Example 2.9: *A strip footing 1.2 m wide is located at a depth of 2 m on non-cohesive soil deposit with average SPT blow count N = 30 (corrected value). Water table is located at a depth of 3 m below ground surface, find allowable bearing pressure for the soil. (Use Terzaghi formula).*

Solution : For strip footing.

$$q_{nu} = 0.471 \, N^2 \, BW_\gamma + 0.785 \, (100 + N^2) \, d \cdot W_q$$

As water table is at large depth, so $W_q = W_{\gamma} = 1.0$.

$$q_{nu} = 0.471 \, (30)^2 \, (1.2) \, (1.0)$$
$$+ \, 0.785 \, (100 + 30^2) \, (2) \, (1)$$
$$= 2078.68 \text{ kN/m}^2$$

Net safe bearing capacity,

$$q_{ns} = \frac{q_{nu}}{F}$$

$$= \frac{2078.68}{3.0}$$

$$= 692.89 \text{ kN/m}^2$$

Example 2.10: *Determine the safe bearing capacity of a footing in purely clayey soil having unit weight 20kN/m³ and cohesion of 30kN/m². If footing is placed at a depth of 1m. Assume soil fail by general shear. By using Terzaghi /IS code/ Skempton/ Meyerhof/ Size of footing is 2m.*

If footing is strip/square/circular/rectangular (L = 2.5m)

Solution: For clayey soil $\varphi = 0$, $K_p = 1$

(i) Terzaghi Method: for $\varphi = 0$, bearing capacity factors

$N_c = 5.7$, $N_q = 1$ and $N_\gamma = 0$

Shape factors: strip ($S_c = 1$, $S_q = 1$ and $S_\gamma = 1$),

square $S_c = 1.3$, $S_q = 1$ and $S_\gamma = 0.8$)

Circular $S_c = 1.3$, $S_q = 1$ and $S_\gamma = 0.6$

rectangular $S_c = 1.0 + 0.3\,(0.8) = 1.24$, $S_q = 1$

and $S_\gamma = 1 - (0.2 \times 0.8) = 0.84$

Shape	q_u	$q_{nu} = q_u - q$	$q_{ns} = q_{nu}/F$	$q_s = q_{ns} + q$
Strip	$30 \times 5.70 \times (1 + 20 \times 1 \times 1) + 0 = 191$	171	57	77
Square	$30 \times 5.7 \times 1.3 + (20 \times 1 \times 1) + 0 = 242.3$	222.3	74.1	94.1
Circular	$30 \times 5.7 \times 1.3 + (20 \times 1 \times 1) + 0 = 242.3$	222.3	74.1	94.1
Rectan.	$30 \times 5.7 \times 1.24 + (20 \times 1 \times 1) + 0 = 232.04$	212.04	70.68	90.68

(ii) IS Code Method:

$$q_{nu} = cN_cS_cd_ci_c + q\,(N_q - 1)\,S_qd_qi_q + 0.5\,\gamma N_\gamma S_\gamma d_\gamma i_\gamma$$

$$B/L = 0.8, \quad D/B = 0.5$$

for $\varphi = 0$,

Bearing Capacity Factors

$$N_c = 5.14, \; N_q = 1 \text{ and } N_\gamma = 0 \Rightarrow N_q - 1 = 0$$

Shape Factors :

Strip ($S_c = 1$, $S_q = 1$ and $S_\gamma = 1$), square $S_c = 1.3$, $S_q = 1.2$ and $S_\gamma = 0.8$.

Circular $S_c = 1.3$, $S_q = 1.2$ and $S_\gamma = 0.6$

Rectangular $S_c = S_q = 1 + 0.2\,(0.8) = 1.16$

$$S_\gamma = 1 - 0.4\,(0.8) = 0.68.$$

Depth Factors:

$$d_c = 1 + 0.2\,(0.5) / (\sqrt{1})$$
$$= 1.1, \; d_q = d_\gamma = 1$$

Inclination Factors: For vertical load all inclination factors = 1.

Shape	q_{nu}	q_{ns} $= q_{nu}/F$	q_s $= q_{ns} + q$
Strip	$30 \times 5.14 \times 1 \times 1.1 \times 1 = 169.62$	56.54	76.54
Square	$30 \times 5.14 \times 1.3 \times 1.1 \times 1 = 220.5$	73.5	93.5
Circular	$30 \times 5.14 \times 1.3 \times 1.1 \times 1 = 220.5$	73.5	93.5
Rectan.	$30 \times 5.14 \times 1.16 \times 1.1 \times 1$ $= 196.76$	65.59	85.59

(iii) Skempton's Method

$$q_{nu} = cN_cN_c$$

$$= 5.14 \times \left(1 + 0.2\frac{B}{L}\right)\left(1 + 0.2\frac{D}{B}\right)$$

$B/L = 0.8$, $D/B = 0.5$

Shape	N_c	q_{nu}	$q_{ns} =$ q_{nu}/F	$q_s =$ $q_{ns} + q$
Strip	$(5.14)(1)(1.1) = 5.94$	30×5.94 $= 178.2$	59.4	79.4
Square	$(5.14)(1.2)(1.1) = 6.78$	30×6.78 $= 203.4$	67.8	87.8
Circular	$(5.14)(1.2)(1.1) = 6.78$	30×6.78 $= 203.4$	67.8	87.8
Rectan.	$(5.14)(1.16)(1.1)$ $=$ 6.56	30×6.56 $= 196.8$	65.6	85.6

(iv) Meyerhoff Method:

$$q_u = cN_cS_cd_ci_c + qN_qS_qd_qi_q + 0.5\gamma BN_\gamma S_\gamma d_\gamma i_\gamma$$

B.C. factors: $N_c = 5.14$, $N_q = 1$ and $N_\gamma = 0$.

Shape Factors:

$$S_c = 1 + 0.2\, K_p\frac{B}{L}$$

and $S_q = S_\gamma = 1$ for $\varphi = 0$,

     Strip $S_c = 1$,

Square/Circular

$$S_c = 1 + (0.2)\,(1)\,(1) = 1.2,$$

Rectangular $S_c = 1 + (0.2)\,(1)\,(0.8) = 1.16$

Depth Factors:

$$d_c = 1 + 0.2\sqrt{K_p}\frac{D}{B} \text{ and}$$

$$d_q = d_\gamma = 1 \text{ for } \varphi = 0^0.$$

$$d_c = 1 + (0.2)\,(1)\,(0.5)$$

$$= 1.1$$

Inclination Factors: for vertical load all inclination factors will be 1

Shape	q_u	q_{nu}	q_{ns}	q_s
Strip	$30(5.14)(1)(1.1)(1)+20(1)(1)(1)(1)$ $= 189.62$	169.62	56.54	76.54
Square	$30(5.14)(1.2)(1.1)(1)+20(1)(1)(1)(1)$ $= 223.54$	203.54	67.84	87.84
Circular	$30(5.14)(1.2)(1.1)(1)+20(1)(1)(1)(1)$ $= 223.54$	203.54	67.84	87.84
Rectan.	$30(5.14)(1.16)(1.1)(1)+20(1)(1)(1)(1)$ $= 196.76$	176.76	58.92	78.92

Example 2.11: *In example 2.10 what will be change in SBC if footing is placed on ground level?*

Solution: when footing is placed at ground level then surcharge will be zero,(q =0), $D/B = 0$, in this case

$$q_{nu} = q_u - q = q_u$$

and $q_s = q_{ns} = \dfrac{q_u}{F}$

(i) Terzaghi $q_u = cN_cS_c = 30\,(5.7)\,S_c = 171\,S_c$

Shape	q_u	q_{nu} $= q_u - q$	q_{ns} $= q_{nu}/F$	q_s $= q_{ns} + q$
Strip	$171(1)$	171	57	57
Square	$171\,(1.3) = 222$	222	74	74
Circular	$171\,(1.3) = 222$	222	74	74
Rectan.	$171\,(1.24) = 212$	212	70.68	70.68

(ii) IS Code Method

as $D/B = 0$ thus $d_c = 1$

$q_{nu} = cN_cS_cd_ci_c = 30\,(5.14)\,S_c = 154.2\,S_c$

Shape	q_{nu}	$q_{ns} = q_{nu}/F$	$q_s = q_{ns} + q$
Strip	$154.2(1) = 154.2$	51.4	51.4
Square	$154.2(1.3) = 200.46$	66.82	66.82
Circular	$154.2(1.3) = 200.46$	66.82	66.82
Rectan.	$154.2(1.16) = 178.87$	59.62	59.62

(iii) Skemptons Method:

$$N_c = 5.14 \times \left(1 + 0.2\frac{B}{L}\right)\left(1 + 0.2\frac{D}{B}\right)$$

$$= 5.14\left(1 + 0.2\frac{B}{L}\right)(1)$$

Shape	Nc	q_{nu}	q_{ns} = q_{nu}/F	q_s = q_{ns} + q
Strip	(5.14)(1)(1.0) = 5.14	30 × 5.14 = 154.2	51.4	51.4
Square	(5.14)(1.2)(1.0) = 6.17	30 × 6.78 = 203.4	67.8	67.8
Circular	(5.14)(1.2)(1.0) = 6.17	30 × 6.78 = 203.4	67.8	67.8
Rectan.	(5.14)(1.16)(1.0) = 5.96	30 × 6.56 = 196.8	65.6	65.6

(iv) Meyerhoff Method:

$$q_u = cN_cS_cd_ci_c$$
$$= 30\,(5.14)\,S_c\,(1)\,(1) = 154.2\,S_c$$

Shape	q_u	q_{nu}	q_{ns}	q_s
Strip	154.2(1) = 154	154	51	51
square	154.2(1.2) = 185	185	61	61
circular	154.2(1.2) = 185	185	61	61
Rectan.	154.2(1.16) = 178	178	59	59

When footing is placed on ground surface?

	Strip			Square			Circular			Rectangular		
	GS	SF	%	GS	SF	%	GS	SF	%	GS	SF	%
Terzaghi	77	57	26	94	74	21	94	74	21	90	70	22
IS code	76	51	33	93	66	29	93	66	29	85	59	31
Skempton	79	51	35	87	67	23	87	67	23	85	65	24
Meyerhoff	76	51	35	87	61	23	87	61	23	78	59	24

Example 2.12 : *Calculate safe bearing capacity of rectangular footing of size 2m × 2.5m placed at a depth of 1m in purely granular soil having* $\gamma = 18kN/m^3$*, c = 0 and* φ *= 30°. For* $\varphi = 30°$ *bearing capacity factors are as given below*

	Terzaghi	IS code	Meyerhof
N_c	37.2	30.14	30.13
N_q	22.5	18.4	18.4
N_γ	19.7	22.4	15.7

Solution: Value of c = 0, thus first term of bearing capacity equation becomes zero. Assume soil fail by general shear

(i) Terzaghi Method:

$$q_u = cN_cS_c + qN_qS_q + 0.5\gamma BN_\gamma S_\gamma$$
$$= qN_qS_q + 0.5\gamma BN_\gamma S_\gamma$$

Bearing capacity factors for $\varphi = 30°$ $N_q = 22.5$ and $N_\gamma = 19.7$ [given in the question]

Shape factor: for rectangular footing $S_q = 1$, and

$$S_\gamma = 1 - 0.2\frac{B}{L} = 0.84$$

Surcharge term q = 18 × 1 = 18, YB term = 18 × 2 = 36

Ultimate bearing capacity

$$q_u = 18\,(22.5)\,(1) + 0.5\,(36)\,(19.7)\,(0.84)$$
$$= 702.86 \text{ kPa}$$

Net ultimate bearing capacity

$$q_{nu} = q_u - q = 702.86 = 684.86$$

Net safe bearing capacity

$$q_{ns} = \frac{q_{nu}}{F} = \frac{684.86}{3} = 228.29$$

Safe bearing capacity

$$q_s = q_{ns} + q = 228.29 = 246.29$$

(ii) IS Code Method

$$q_{nu} = q\,(N_q - 1)\,S_qd_qi_q + 0.5\,BN_\gamma\,S_\gamma d_\gamma i_\gamma$$

Bearing capacity factors for $\varphi = 30°$ $N_q = 18.4$ and $N_\gamma = 22.4$ [given in the question]

Shape factor: for rectangular footing

$$S_q = 1 + 0.2\frac{B}{L}$$

and

$$S_\gamma = 1 - 0.4\frac{B}{L} = 0.68$$

Depth factor

$$d_q = d_\gamma = 1 + 0.1\frac{D}{B}\sqrt{K_p}$$

for $\phi > 10\ \frac{D}{B} = 0.5$ and $K_p = \frac{1 + \sin\varphi}{1 - \sin\varphi} = 3$

$$d_q = d_\gamma = 1 + 0.1\,(0.5)\,\sqrt{3} = 1.09$$

Inclination factors: as load is vertical all inclination factors are 1

Surcharge term q = 18 × 1 , YB term = 18 × 2 = 36

Net ultimate bearing capacity

$$q_{nu} = 18\,(17.4)\,(1.16)\,(1.09)\,(1) + 0.5\,(36)\,(22.4)\,(0.68)\,(1.09)\,(1)$$
$$= 694.86$$

Net safe bearing capacity

$$q_{ns} = \frac{q_{nu}}{F}$$
$$= \frac{694.86}{3} = 231.62$$

Safe bearing capacity

$$q_s = q_{ns} + q = 231.62 + 18$$
$$= 249.62$$

(iii) Skempton's Method:

This method is applicable only for purely cohesive soil. So this method is not applicable for this problem

(iv) Meyerhof's Method:

$$q_u = qN_qS_qd_qi_q + 0.5\gamma BN_\gamma S_\gamma d_\gamma i_\gamma$$

Bearing capacity factors for $\varphi = 30°$ $N_q = 18.4$ and $N_\gamma = 15.7$ [given in the question]

Shape Factor: for rectangular footing for $\varphi > 10$

$$S_q = S_\gamma = 1 + 0.1\, K_p \frac{B}{L}$$

$$= 1 + 0.1\,(3)\,(0.8) = 1.24$$

Depth Factor

For $\varphi > 10$ $d_q = d_\gamma = 1 + 0.1 \frac{D}{B}\sqrt{K_p}\, d_q$

$$= d_\gamma = 1 + 0.1\,(0.5)\sqrt{3} = 1.09$$

Inclination Factors: All inclination factors will be 1

Surcharge term

$$q = 18 \times 1 = 18$$

$$\text{YB term} = 18 \times 2 = 36$$

Ultimate bearing capacity

$q_u = 18\,(18.4)\,(1.24)\,(1.09)\,1 + 0.5\,(36)\,(15.7)\,(1.24)\,(1.09)\,1$

$$= 829.6 \text{ kPa}$$

Net ultimate bearing capacity

$$q_{nu} = q_u - q$$

$$= 829.6 - 18 = 811.6$$

Net safe bearing capacity

$$q_{ns} = \frac{q_{nu}}{F} = \frac{811.6}{3} = 270.73$$

Safe bearing capacity

$$q_s = q_{ns} + q$$

$$= 270.53 + 18 = 288.53$$

Example 2.13 : *Design a footing to carry a load of 3000kN to be placed at a depth of 1m in a soil having the following properties having $\gamma = 20$ kN/m³, c = 25 kPa and $\varphi = 30^0$. Assume soil fail by general shear and water table is at large depth. Bearing capacity factors for $\varphi = 30^0$ are as given below*

	Terzaghi	IS code	Meyerhof
N_c	37.2	30.14	30.13
N_q	22.5	18.4	18.4
N_γ	19.7	22.4	15.7

Solution:

$$q_s = q_{ns} + q = \frac{q_{nu}}{F} + q$$

$$= \frac{q_u - q}{F} + q = \frac{1}{F}[q_u + q\,(F - 1)]$$

$$= 0.33\, q_u + 0.67\, q$$

$$P = q_sA = (0.33\, q_u + 0.67\, q)\, A$$

i.e. $3000 = (0.33\, q_u + 0.67\, q)\, A$

(i) Terzaghi Method:

$$q_u = cN_cS_c + qN_qS_q + 0.5\,\gamma BN_\gamma S_\gamma$$

$$= 25(37.2)S_c + 20(22.5)\,S_q + 0.5\,(20B)(19.7)\,S_\gamma$$

$$q_u = 930S_c + 450S_q + 197BS_\gamma$$

shape factors for various shapes are as below

Strip Footing :

$$S_c = S_q = S_\gamma = 1$$

thus $q_u = 930 + 450 + 197\,B = 1380 + 197\,B$

$$3000 = (0.33\, q_u + 0.67\, q)\, A$$

$$= [0.33\,(1380 + 197B) + 0.67(20)\,[BX1]$$

$$3000 = 468.8\,B + 65.01\,B^2$$

$$\therefore \quad B = 4.08\text{m}$$

Square Footing:

$$S_c = 1.3\,, S_q = 1 \text{ and } S_\gamma = 0.8$$

thus $q_u = 930\,(1.3) + 450\,(1) + 197\,B\,(0.8)$

$$= 1659 + 157.6B$$

$$3000 = (0.33\, q_u + 0.67\, q)\, A$$

$$= [0.33\,(1659 + 157.6B + 0.67\,(20)]B^2$$

$$3000 = 560.87\,B^2 + 52\,B^3$$

$$\therefore \quad B = 2.11\text{m}$$

Circular Footing :

$$S_c = 1.3\,, S_q = 1 \text{ and } S_\gamma = 0.6$$

thus $q_u = 930\,(1.3) + 450\,(1) + 197B\,(0.6)$

$$= 1659 + 118.2B$$

thus $0.33\, q_u = 547.5 + 39B$

$0.33\, q_u + 0.67\, q = 547.5 + 39B + 0.67\,(20)$

$$= 561 + 39B$$

$$3000 = (0.33q_u + 0.67q)\, A$$

$$= [561 + 39B] \times 0.785\,B^2$$

$$3000 = 30.61\,B^3 + 440.38\,B^2$$

$$\therefore \quad B = 2.41\text{m}$$

(d) Rectangular footing (assuming L/B = 1.25)

$$S_c = 1 + 0.3\,(0.8) = 1.24, \; S_q = 1$$

and $\quad S_\gamma = 1 - 0.2\,(0.8) = 0.84$

$\quad q_u = 930\,(1.24) + 450\,(1) + 197\,B\,(0.84)$

$\quad\quad = 1603.2 + 165.48\,B$

thus $\quad 0.33\,q_u = 529.05 + 54.61\,B$

$0.33\,q_u + 0.67\,q = 529.05 + 54.61\,B + 0.67\,(20)$

$\quad\quad = 542.45 + 54.61\,B$

$\quad 3000 = (0.33\,q_u + 0.67\,q)\,A$

$\quad\quad = [542.45 + 54.61\,B]\,[1.25B^2]$

$\quad 3000 = 68.26\,B^3 + 678.06\,B^2$

$\therefore \quad B = 1.92\text{m}$

and $\quad L = 1.25B = 2.4\text{m}$

(ii) IS Code Method:

$$q_s = q_{ns} + q = \frac{q_{nu}}{F} + q$$

$$= 0.33\,q_{nu} + 20$$

$$q_{nu} = cN_cS_cd_ci_c + q\,(N_q - 1)S_qd_qi_q + 0.5\,\gamma BN_\gamma S_\gamma d_\gamma i_\Gamma$$

Depth factors $d_c = 1 + 0.2\,(D/B)\sqrt{K_p}$

$$= 1 + 0.2\,(1/B)\,(\sqrt{3}) = 1 + \frac{0.35}{B}$$

$$d_q = d_\gamma = 1 + 0.1\,(D/B)\sqrt{Kp}$$

$$= 1 + 0.1\,(1/B)\,(\sqrt{3})$$

$$= 1 + \frac{0.17}{B}$$

$$= 25(30.14)\,S_c\left[1 + \frac{0.35}{B}\right]$$

$$+ 20(17.4)\,S_q\left[q + \frac{0.17}{B}\right]$$

$$+ 0.5(20B)(22.4)\,S_\gamma\left[1 + \frac{0.17}{B}\right]$$

$$q_{nu} = 753.5\,S_c\left[1 + \frac{0.35}{B}\right] + 348\,S_q\left[1 + \frac{0.17}{B}\right]$$

$$+ 224\,BS_\gamma\left[1 + \frac{0.17}{B}\right]$$

$$q_s = 248.6\,S_c\left[1 + \frac{0.35}{B}\right] + 114.84\,S_q$$

$$\left[1 + \frac{0.17}{B}\right] + 73.92\,BS_\gamma\left[1 + \frac{0.17}{B}\right] + 20$$

$$P = q_sA$$

Strip Footing :

$$S_c = S_q = S_\gamma = 1$$

thus $\quad q_u = 930 + 450 + 197\,B$

$$= 1380 + 197\,B$$

$$q_s = 248.6\left[1 + \frac{0.35}{B}\right] + 114.84\left[1 + \frac{0.17}{B}\right]$$

$$+ 73.92\,B\left[1 + \frac{0.17}{B}\right] + 20$$

$$= 396 + 73.92\,B + \frac{106.53}{B}$$

$$3000 = \left(396 + 73.92\,B + \frac{106.53}{B}\right)(B \times 1)$$

$\therefore \quad 73.92\,B^2 + 396B + 106.53 = 3000$

$\quad 73.92\,B^2 + 396\,B - 2893.47 = 0$

$\quad\quad B = 4.13\text{m}$

Square Footing:

$$S_c = 1.3,\ S_q = 1.2\ \text{and}\ S_\gamma = 0.8$$

thus $\quad q_s = 248.6\,(1.3)\left[1 + \frac{0.35}{B}\right] + 114.84\,(1.2)$

$$\left[1 + \frac{0.17}{B}\right] + 73.92\,B\,(0.8)\left[1 + \frac{0.17}{B}\right] + 20$$

$$q_s = 491.04 + \frac{136.54}{B} + 59.14\,B$$

$\therefore \quad 3000 = \left[491.04 + \frac{136.54}{B} + 59.14\,B\right]B^2$

$\quad 3000 = 59.14\,B^3 + 491.04\,B^2 + 136.54B$

$\therefore \quad\quad B = 2.1\text{m}$

Circular footing

$$S_c = 1.3,\ S_q = 1.2\ \text{and}\ S_\gamma = 0.6.$$

$$q_s = 248.6\,(1.3)\left[1 + \frac{0.35}{B}\right] + 114.84\,(1.2)$$

$$\left[1 + \frac{0.17}{B}\right] + 73.92B\,(0.6)\left[1 + \frac{0.17}{B}\right] + 20$$

$$q_s = 488.53 + \frac{136.54}{B} + 44.35\,B$$

$\therefore \quad 3000 = \left[488.53 + \frac{136.54}{B} + 44.35\,B\right]0.785\,B^2$

$\quad 3000 = 34.81\,B^3 + 383.49\,B^2 + 107.18\,B$

$\therefore \quad\quad B = 2.42\text{m}$

Rectangular footing (assuming $L/B = 1.25$)

$$S_c = S_q = 1 + 0.2\,(0.8) = 1.16$$

and $\quad S_\gamma = 1 - 0.4\,(0.8) = 0.68$

$$q_s = 248.6\,(1.16)\left[1 + \frac{0.35}{B}\right] + 114.84\,(1.16)$$

$$\left[1 + \frac{0.17}{B}\right] + 73.92B\,(0.68)\left[1 + \frac{0.17}{B}\right] + 20$$

$$q_s = \left[450.13 + \frac{123.58}{B} + 50.26\,B\right]$$

$\therefore \quad q_s = \left[450.13 + \frac{123.58}{B} + 50.26\,B\right]$

$\therefore \quad 3000 = \left[450.13 + \dfrac{123.58}{B} + 50.26\,B\right]1.25\,B^2$

$3000 = 62.82\,B^2 + 562.66\,B^2 + 154.47\,B$

$\therefore \quad B = 1.98m$ and $L = 1.25B = 2.47m$

(iii) Meyerhof Method :

$q_s = q_{ns} + q = \dfrac{q_{nu}}{F} + q$

$\quad = \dfrac{q_u - q}{F} + q$

$\quad = \dfrac{1}{F}\,[q_u + q\,(F - 1)]$

$\quad = 0.33\,q_u + 0.67\,qP = q_s A$

$q_s A = (0.33\,q_u + 0.67\,q)\,A$

i.e. $\quad 3000 = (0.33\,q_u + 0.67\,q)\,A$

$q_u = cN_cS_cd_ci_c + qN_qS_qd_qi_q + 0.5\gamma BN_\gamma S_\gamma d_\gamma i_\gamma$

Inclination factors $i_c = i_q = i_\gamma = 1$

Depth factors $d_c = 1 + 0.2\,(D/B)\,\sqrt{K_p}$

$\quad = 1 + 0.2\,(1/B)\,(\sqrt{3}) = 1 + \dfrac{0.35}{B}$

$d_q = d_\gamma = 1 + 0.1\,(D/B)\,\sqrt{K_p}$

$\quad = 1 + 0.1\,(1/B)\,(\sqrt{3})$

$\quad = 1 + \dfrac{0.17}{B}$

$q_u = 25\,(30.13)$

$\quad S_c\left[1 + \dfrac{0.35}{B}\right] + 20\,(18.4)\,S_q\left[1 + \dfrac{0.17}{B}\right]$

$\quad + 0.5\,(20)\,B\,(15.7)\,S_\gamma\left[1 + \dfrac{0.17}{B}\right]q_u$

$\quad = [753.25\,S_c + 368\,S_q + 26.69\,S_\gamma]$

$\quad + \dfrac{1}{B}\,[263.64\,S_c + 62.56\,S_q] + 157\,BS_\gamma$

$q_s = 0.33\,q_u + 0.67q$

$\quad = [248.57\,S_c + 121.44\,S_q + 8.81\,S_\gamma]$

$\quad + \dfrac{1}{B}\,[87S_c + 20.64\,S_q] + 157BS_\gamma + 13.4$

Strip Footing :

$S_c = 1 + 0.2\,K_p\dfrac{B}{L}$

$S_q = S_\gamma = 1 + 0.1\,K_p\dfrac{B}{L}$

thus $\quad S_C = 1,\ S_q = S_\gamma = 1$

$q_s = [248.57\,(1) + 121.44\,(1) + 8.81\,(1)$

$\quad + \dfrac{1}{B}\,[87\,(1) + 20.64\,(1)] + 157(1) + 13.4$

$q_s = 392.22 + 157\,B\,\dfrac{107.64}{B}$

$3000 = \left[392.22 + 157\,B + \dfrac{107.64}{B}\right](B \times 1)$

$\therefore \quad 157B^2 + 392.22\,B + 107.64 = 3000$

$157B^2 + 392.22\,B - 2892.36 = 0$

$\therefore \quad B = 3.22$

Square Footing:

$S_c = 1 + 0.2\,K_p\dfrac{B}{L}$

$S_q = S_\gamma = 1 + 0.1\,K_p\dfrac{B}{L}$

thus $S_c = 1.6,\ S_q = S_\gamma = 1.3$

$q_s = [248.57\,(1.6) + 121.44\,(1.3) + 8.81\,(1.3)$

$\quad + \dfrac{1}{B}\,[87\,(1.6) + 20.64(1.3)] + 157B(1.3) + 13.4$

$q_s = 567.04 + \dfrac{166.03}{B} + 204.1\,B$

$3000 = \left[567.04 + \dfrac{166.03}{B} + 204.1\,B\right]B^2$

$\therefore \quad 204.1\,B^3 + 567.04\,B^2 + 166.03\,B = 3000$

$\therefore \quad B = 1.72m$

Circular Footing

$S_c = 1 + 0.2\,K_p\dfrac{B}{L}$

$S_q = S_\gamma = 1 + 0.1\,K_p\dfrac{B}{L}$

thus $S_c = 1.6,\ S_q = S_\gamma = 1.3$

$q_s = [248.57(1.6) + 121.44\,(1.3) + 8.81\,(1.3)$

$\quad + \dfrac{1}{B}[87\,(1.6) + 20.64\,(1.3)]$

$\quad + 157\,B(1.3) + 13.4$

$q_s = 567.04 + \dfrac{166.03}{B} + 204.1\,B$

$\left[567.04 + \dfrac{166.03}{B} + 204.1\,B\right]0.785\,B^2$

$\quad = 3000$

$\therefore \quad 160.02\,B^3 + 445.13\,B^2 + 130.33\,B = 3000$

$\therefore \quad B = 1.91\,m$

Rectangular footing (assuming L/B = 1.25)

$S_c = 1 + 0.2\,K_p\dfrac{B}{L},\ S_q = S_\gamma = 1 + 0.1\,K_p\dfrac{B}{L}$

thus

$$S_c = 1.48 \,,\, S_q = S_\gamma = 1.24$$

$$q_s = [248.57(1.48) + 121.44(1.24) + 8.81\,(1.24)$$

$$+ \frac{1}{B}[87\,(1.48) + 20.64\,(1.24)]$$

$$+ 157\,B(1.24) + 13.4$$

$$q_s = \left[529.39 + \frac{154.35}{B} + 13.4 + 194.68\,B \right]$$

$$= \left[194.68\,B + \frac{154.35}{B} + 542.79 \right]$$

$$3000 = \left[194.68\,B + \frac{154.35}{B} + 542.79 \right] (1.25\,B^2)$$

$$\therefore \quad 243.35\,B^3 + 678.49\,B^2 + 192.94\,B = 3000$$

$$\therefore \quad\quad B = 1.59\,M$$

$$B = 1.59\,m \quad L = 1.99\,m$$

Example 2.14: *Determine immediate settlement of square footing of 1m size founded at a depth of 1m in a soil with E = 10000kN/m², μ = 0.3. The footing is subjected to a pressure of 200kN/m². Assume influence coefficient is 1.*

Solution : Immediate settlement is given by

$$S_i = qB \left[\frac{1 - \mu^2}{E} \right] I$$

$$= 200 \times 1 \times \left[\frac{1 - 0.3^2}{10000} \right]$$

$$= 18.2\,mm$$

Example 2.15: *A square footing on sand at 2 m depth shows an elastic settlement of 5.5 mm. Under a loading of 200 kN/m². How much a footing would settle if it has to carry a load of 150 kN/m².*

Solution: Elastic settlement is given by,

$$S = q \cdot B \left(\frac{1 - \mu^2}{E_S} \right) \cdot I$$

As size of footing remains same i.e. B is constant.

As sub-soil is same its parameter remains same (μ, E_s and I remain constant)

$$S \propto q$$

$$\therefore \quad \frac{S}{q} = constant$$

$$S_2 = S_1 \left(\frac{q_2}{q_1} \right)$$

$$\therefore \quad S_2 = 5.5 \left[\frac{150}{200} \right]$$

$$= 4.125\,mm$$

Example 2.16: *For a clayey soil what will be the settlement for rigid rectangular footing 2 m × 3 m if elastic parameters E_s = 8000 kN/m² , μ = 0.35, I_s = 0.785 and μ = 0.35 under a load of 600 kN ? How the settlement is affected if footing is flexible ?*

Solution :
$$S = q \cdot B \left[\frac{1 - \mu^2}{E_S} \right] \cdot I$$

$$= 100\,(2) \left[\frac{1 - 0.35^2}{8000} \right] 0.875$$

$$= 0.018\,m$$

For flexible footing,

$$S = q \cdot B \left[\frac{1 - \mu^2}{E_S} \right] I = 0.022\,I$$

For
$$\frac{L}{B} = \frac{3}{2} = 1.5$$

Influence factor I = 1.20

$$\therefore \quad S = (0.022)\,(1.20)$$

$$= 0.0264\,m$$

Settlement in case of flexible footing is more as compared to rigid footing.

Example 2.17: *Calculate settlement of plate if permissible settlement of foundation is to be limited to 40 mm. Assume width of plate and that of foundation 300 mm and 2000 mm respectively.*

Solution : Settlement is given by

$$S = q \cdot B \left[\frac{1 - \mu^2}{E_S} \right] I$$

For given intensity of loading and soil $q \left[\dfrac{1 - \mu^2}{E_S} \right] I$ is constant

$$\therefore \quad S \propto B$$

$$\therefore \quad \frac{S}{B} = Constant$$

$$\therefore \quad S_p = \left(\frac{S}{B} \right)_{footing} \cdot B_{plate}$$

$$\therefore \quad S_p = \left(\frac{40}{2000} \right) 300 = 6\,mm$$

Example 2.18 : *The overburden pressure at the middle of 7.5 cm thick clayey layer increased from 2 kg/cm² to 3.5 kg/cm². Find settlement due to consolidation assuming liquid limit and initial void ratio of clay as 36% and 0.82 respectively.*

Solution : Settlement is given by,

$$S = C_c \cdot \frac{H}{1+e} \cdot \log\left(\frac{p + \Delta p}{p}\right)$$

Here, $C_c = 0.009\,(\omega_L - 10)$

$$= 0.009\,(36 - 10)$$

$$= 0.234 \ \{\text{Assuming undisturbed soil}\}$$

$$\therefore \quad S = 0.234\,\frac{7.5}{1 + 0.82}\,\log\left[\frac{3.5}{2.0}\right] = 0.2343 \text{ cm}$$

Example 2.19 : Determine consolidation settlement of clay layer 1.5 m thick at a depth of   15 m below ground level. If water table is lowered from ground level to 1.0 m depth. The clay layer underlays the sand bed. The soil properties are as under.

Sand : $\gamma_{moist} = 17.5$ kN/m^2, $\gamma_b = 10$ kN/m^3, $G = 2.65$

Clay : $\omega = 35\%$, $\omega_L = 45\%$, $\gamma_b = 9$ kN/m^3, $G = 2.70$.

Solution : For sand : $\gamma_b = 10$ kN/m^3, $\gamma = 17.5$ kN/m^3

For clay : $\gamma_b = 9$ kN/m^3.

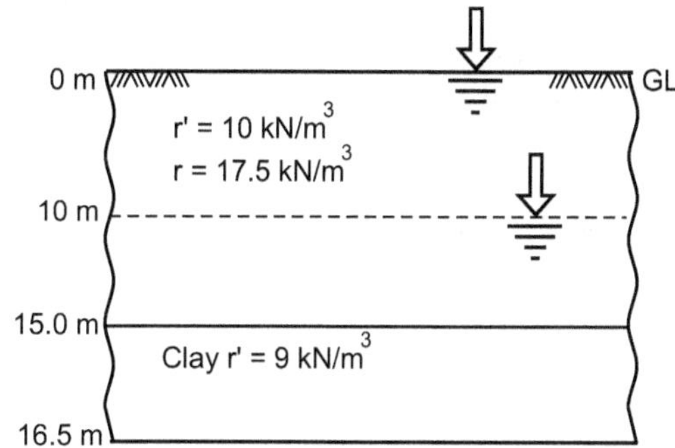

Fig. 2.10

Vertical stress at centre of clay layer :

(i)   Prior to lowering water table

$$\sigma_0' = (10)(15) + 9(0.75) = 156.75 \text{ kN/m}^2$$

(ii)   After lowering of water table

$$\sigma_1' = (17.5)(10) + (10)(5) + 9(0.75)$$

$$= 231.75 \text{ kN/m}^2$$

For clay $C_c = 0.009\,(\omega_L - 10) = 0.315$

and $e_o = \dfrac{\omega G}{S_\gamma} = \dfrac{(0.35)(2.70)}{1.0} = 0.945$

Consolidation settlement of clay layer

$$S = \frac{C_c H}{1 + e}\,\log_{10}\left(\frac{\sigma_0' + \Delta\sigma'}{\sigma_0'}\right)$$

$$\therefore \quad S = \frac{(0.315)(1.5)}{1 + 0.945}\,\log_{10}\left(\frac{231.75}{156.75}\right)$$

$$\therefore \quad S = 0.041 \text{ m}$$

Example 2.20 : A raft 6 m × 10 m is supported on sand in above problem at a depth of 3 m and carries a gross loading of 100 kN/m^2. Determine the consolidation settlement under the load accounting for floating effect. Assume water table at ground surface.

Solution : Referring Fig. 2.11.

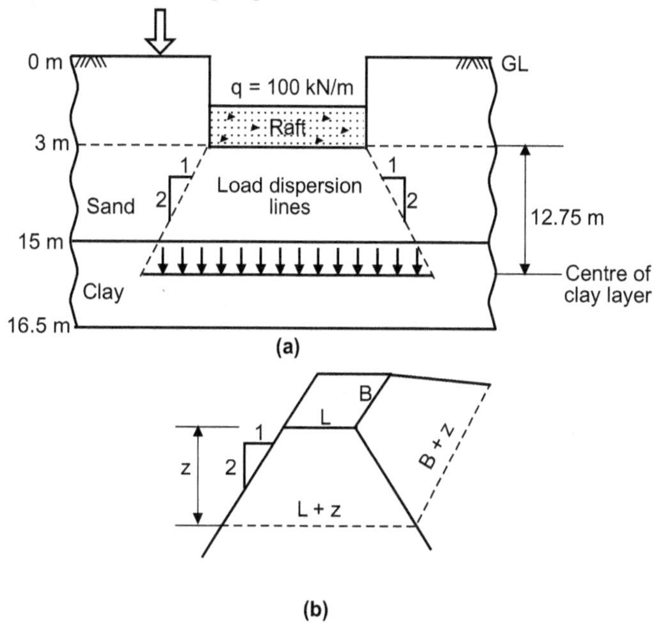

Fig. 2.11

(i)   Initial overburden pressure at centre of clay layer

$$\sigma_0' = (10)(15) + 9(0.75)$$

$$= 156.75 \text{ kN/m}^2$$

(ii)   Calculation of increase in pressure at centre of clay layer due to footing.

Gross load due to footing = $(100) \times (6 \times 10) = 6000$ kN.

Weight of excavated soil = $(10)(3)(6 \times 10) = 1800$ kN

Net downward load at base of footing = $Q = 4200$ kN

Assuming load is dispersed in ratio 1 : 2

$$\Delta\sigma' = \frac{Q}{(10 + 12.75)(6 + 12.75)}$$

$$= \frac{4200}{426.5625}$$

$$= 9.85 \text{ kN/m}^2$$

Settlement of footing,

$$S = \frac{C_c H}{1 + e}\,\log_{10}\left(\frac{\sigma_0' + \Delta\sigma'}{\sigma_0'}\right)$$

$$\therefore \quad S = \frac{(0.315)(1.5)}{(1 + 0.945)}\,\log_{10}\left(\frac{156.75 + 9.85}{156.75}\right)$$

$$\therefore \quad S = 6.43 \times 10^{-3} \text{ m}$$

Example 2.21 : *Determine consolidation settlement of a raft 10 m × 15 m loaded at a depth of 3 m (on sand) carrying a net load of 15000 kN. If a compressible clay layer is found to exist sandwiched between two sandy layers 10 m beneath the raft, given following particulars.*

Thickness = 2.0 m, ω_n = 35%, ω_L = 55, γ_b = 9 kN/m^3 (both for sand and clay), G = 2.7.

Solution :

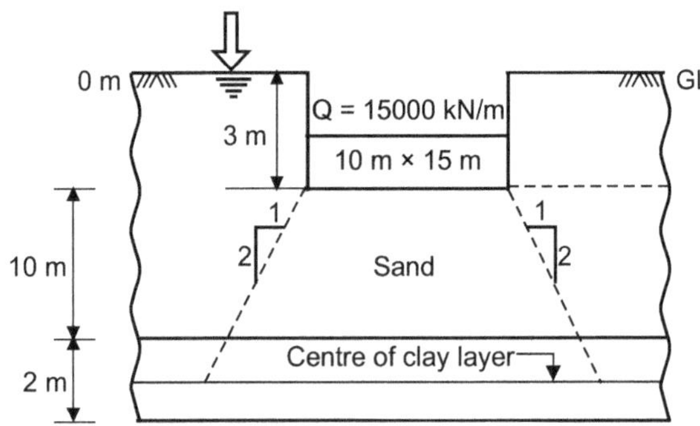

Fig. 2.12

Assuming water table at ground surface.

Initial overburden pressure at the centre of clay layer,

$$\sigma_0' = 9\,(13) + 9\,(1) = 144 \text{ kN/m}^2$$

Increase in pressure,

$$\Delta\sigma' = \frac{Q}{(B + z)\,(L + z)} = \frac{Q}{(10 + 11)\,(15 + 11)}$$

$$= \frac{15000}{21 \times 26} = 27.47 \text{ kN/m}^2$$

Settlement,

$$S = \frac{\sigma_c H}{1 + e}\, \log_{10}\!\left(\frac{\sigma_0' + \Delta\sigma'}{\sigma_0'}\right)$$

For clay : C_c = 0.009 (ω_L – 10)

$$= 0.009 \times 45 = 0.405$$

and $\quad e = \dfrac{\omega G}{S_\gamma} = \dfrac{0.35 \times 2.7}{1.0} = 0.945$

$\therefore \quad S = \dfrac{(0.405)\,(2.0)}{(1 + 0.945)}\, \log_{10}\!\left(\dfrac{144 + 27.47}{144}\right)$

$$S = 0.032 \text{ m}$$

EXERCISE

1. Distinguish between local shear failure and general shear failure.

2. What are basic characteristics of failure mechanism in general shear and local shear failure ?

3. Sketch load settlement curve to show local shear, general shear and punching shear failure.

4. Draw suitable sketches to illustrate failure surface and load settlement curve for the following :

   (i)   Local shear failure

   (ii) General shear failure

   (iii) Punching shear failure

5. 'Some of assumptions made by Terzaghi while deriving equation for bearing capacity of shallow foundation are not valid'. Offer your comments on these invalid assumptions.

6. State and explain Terzaghi's equation of bearing capacity.

7. What are the assumptions made in Terzagh's analysis of bearing capacity of continuous footing ?

8. State and explain UBC equation for strip footing as suggested by Terzaghi and how the same was modified by Hansen for various Parameters.

9. Write short note on Mayerhoff's theory.

10. Write short note on Skempton's bearing capacity theory.

11. Compare with neat sketches bearing capacity equations as suggested by Terzaghi and Mayerhoff and point out how and why modifications were needed ?

12. Explain Skempton's analysis of determination of bearing capacity of Clayey soil.

13. Write short notes on :

   (i)   Bearing capacity of layered soils.

   (ii) Brinch Hansen method for finding bearing capacity.

14. What are the factors influencing the bearing capacity of a footing on a cohesive soil?

15. What are the basic characteristics of failure mechanisms in general shear failure and local shear failure.

16. Explain the effect of submergence of bearing capacity for different positions of ground water table.

17. Write short note on :

   (i)   Skempton's bearing capacity equation.

   (ii) Presumptive bearing capacity of soil.

18. Draw a neat sketch of plate load test. Show thereon ten components and indicate their function. What observations would you collect during the test ?

19. Write short note on :

    (i) Effect of eccentric loading on bearing capacity.

20. State and explain U.B.C. equation for strip footing as suggested by Terzagi and how the same was modified by Hansen's for various parameters.

21. What is elastic settlement ? Explain how it is evaluated.

22. What is contact pressure ? What are the factors on which it depends ? Draw contract pressure distribution diagram for rigid footing.

23. Distinguish between Uniform and non-uniform settlement ?

24. Enlist the causes of differential settlement and explain how to minimise it.

25. What is the effect of lowering of the water table on settlement ?

26. What is angular distortion ? What is its effect on the structural member.

PROBLEMS FOR PRACTICE

1. Compute the safe bearing capacity of square footing 1.5 m × 1.5 m located at a depth of 1 m below the ground level in a sandy soil of average density 20 kN/m^3, $\phi = 20°$, $N_c = 17.7$, $N_q = 7.4$ and $N_\gamma = 5.0$. Assume factor of safety = 3 and water table is very deep. Also compute the reduction in safe bearing capacity of footing if water table rises to the ground level.

   (**Ans. :** 83 kN/m^2, 48 kN/m^2 by 42%)

2. Determine the safe bearing capacity of a circular footing for the following details :

   Diameter = 2m, Depth of foundation = 1.5 m

   Shear strength of soil (S_u) = 25 kN/m^2. Unit weight of soil = 20 kN/m^3

   Value of $N_c = 5.7$, $N_q = 1.0$, $N_\gamma = 0.0$ Factor of safety = 3.0

   (**Ans. :** 60.875 kN/m^2)

3. The construction of a strip-footing is undertaken during a summer period and water table was observed at 2.5 m from ground surface. The soil parameters are density = 19.2 kN/m^3 and $\phi = 32°$ ($N_q = 23.2$, $N_\gamma = 30.2$), c = 0. Width of footing is 3.0 m and depth is 2.0 m. During monsoon water table rises to ground surface. Determine the gross safe bearing capacity in both the cases for a factor of safety = 2.5. Use Terzaghi's water table correction factors.

   (**Ans. :** 582.22 kN/m^2, 375.17 kN/m^2)

4. Determine the ultimate and net bearing capacity, use following data:

   (i) Footing size = 2 m × 2 m, depth of foundation = 1.5 m.

   (ii) Soil density = 1800 kg/m^3, c = 15 kN/m^2, $\phi = 15°$.

   (iii) $N_c' = 9.7$, $N_q' = 2.7$, $N_\gamma' = 0.9$.

   (iv) Use local shear concept.

   (**Ans. :** 211.96 kN/m^2, 184.96 kN/m^2)

5. Determine settlement of footing 2 m × 3 m carrying a load of 1200 kN, if a plate load test (0.3 m square plate) on same soil gave the settlement of 3 mm under a loading of 25 kN.

   (**Ans.** 14.4 mm clayey soil)

6. Determine immediate settlement at the centre of foundation. Use the following data:

   (a) Circular foundation = 20 m diameter.

   (b) Contact pressure = 30 kN/m^2.

   (c) E = 20,000 kN/m^2, $\mu = 0.45$, $\gamma = 22$ kN/m^3

   (d) Influence factor at centre = 1.0 (**Ans. :** 0.024 m)

7. Determine the elastic settlement of a footing 3m × 3m resting on the sandy soil with E = 45,000 kN/m^2, $\mu = 0.3$, $I_s = 0.82$ at centre (rigid) if footing carries a load of 2,000 kN. (**Ans. :** 0.011 m)

8. Estimate average immediate settlement. Use the following data :

   (a) Footing = 4m × 2m

   (b) Depth of foundation = 2m

   (c) E = 48 MN/m^2, v = 0.5

   (d) Contact pressure = 200 kN/m^2

   (e) $\mu_o = 0.78$ and $\mu_1 = 0.84$ (**Ans. :** 4.095×10^{-3} m)

9. For a clayey soil, what will be the total settlement for rigid rectangular footing 2m × 3m if elastic parameters are E = 8000 kN/m^2, μ = 0.35, I_s = 0.785, under a load of 600 kN ? How will the settlement be affected if footing is flexible ?

   (**Ans. :** 0.0196 m)

10. Compressible clayey layer 2 m thick underlies a thick sand bed which carries a footing 2 m × 3 m with an allowable soil pressure of 250 kN/m^2 at a depth of 1.0 m. Examine the consolidation settlement due to the compressible layer at a depth of 10 m below ground surface.

    (**Given :** Clayey layer : γ_b = 10 kN/m^2, ω_L = 50%, ω_n = 35%, Sand layer : γ_b = 9 kN/m^3.

    Assume G = 2.7.

    (**Ans.** 5.023 × 10^{-3} m)

11. Determine the settlement of a compressible layer 2.5 m thick if it is subjected to an effective consolidation pressure of 10 kN/m^2. Its initial overburden pressure corresponding to a void ratio of e_0 = 0.98 is 108 kN/m^2, liquid limit of soil is 45%.

    (**Ans.** 0.169 m)

FOUNDATION FOR DIFFICULT SOILS

3.1 INTRODUCTION

The soil which impose problems during construction of foundation or which impair functioning of foundation and cause damage to the foundation and hence the structure supported by foundation is called difficult soil.

Difficult Soils are Grouped into Following Categories :

1. Weak and compressible soil
2. Collapsible soil
3. Expansive soil
4. Corrosive soil

3.2 WEAK AND COMPRESSIBLE SOIL

The soil which is incapable of supporting the load imposed on it due to its compressibility and following are the soils which belong to this category

- **Clays/Silts/Peats :**
  - ➤ These types of soil deposits are often found near the mouths of rivers, along the perimeters of bays, and beneath swamps or lagoons.
  - ➤ Soil deposits with high organic content are often found in these low lying types of locations and can be especially troublesome.
  - ➤ Since land features in which these troublesome soils are typically found are low lying, they are prone to flooding.
  - ➤ Hence before buildings or roadways can be constructed on such soil deposits, the grade level must be raised by adding compacted fill.
  - ➤ However, adding significant amounts of compacted fill puts significant loads on the soil which can cause significant settlements.
- **Loose Saturated Sands :** Loose saturated sand deposits that are located in seismically active regions are prone to liquefaction and settlements during strong ground motion.

3.2.1 Guidelines for Construction Over Weak and Compressible Soils

Soil Improvement among the various strategies used when encountering extremely weak or compressible soil layers are the following:

- **Removal and Replacement :** This method can be employed when:

  - ➤ The poor soil deposit is relatively small;
  - ➤ The groundwater level is relatively deep;
  - ➤ Good fill soil is readily available.
- **Temporary Surcharge Fills :**
  - ➤ The idea here is to preload the weak/ compressible soil with a temporary surcharge.
  - ➤ The underlying weak/ compressible soil is allowed to consolidate under the surcharge (again sand drains accelerate the process).
  - ➤ The surcharge is removed before the proposed building construction occurs.
  - ➤ Since the building is constructed on over-consolidated soil the displacements are considerably reduced.
- **Vibrocompaction :** This is particularly effective for loose sandy soils.
- **Chemical Stabilization :**
  - ➤ In the past, the weak clays and silts were often mixed with lime and the existing soil pore fluid to cement the soil grains together, making the soil stronger and less compressible.
  - ➤ Presently, the trend in geotechnical engineering is away from using lime and toward using Pulverized Fly Ash (PFA), which is a processed waste product from coal fired electric power generating plants.
  - ➤ Again, the effect is to cement the soil grains together, increasing the soil strength and reducing both its compressibility potential expansivity.

3.3 COLLAPSIBLE SOILS

- Collapsible soils are those that appear to be strong and stable in their natural (dry) state, but which rapidly consolidate under wetting, generating large and often unexpected settlements.
- This can yield disastrous consequences for structures unwittingly built on such deposits. Such soils are often termed "collapsible" or "metastable" and the process of their collapsing is often called any of "hydro-consolidation", "hydro compression", or "hydro-collapse."
- "Loess" which is recognized as potentially collapsible.

- Collapsible soil deposits share two main features:
  1. They are loose, cemented deposits;
  2. They are naturally quite dry.
- Loess soils consist primarily of silt sized particles loosely arranged in a cemented honeycombed structure. The loose structure is held together by small amounts of water softening or water soluble cementing agents such as clay minerals and $CaCO_3$.
- The introduction of water dissolves or softens the bonds between the silt particles and allows them to take a denser packing under any type of compressive loading.

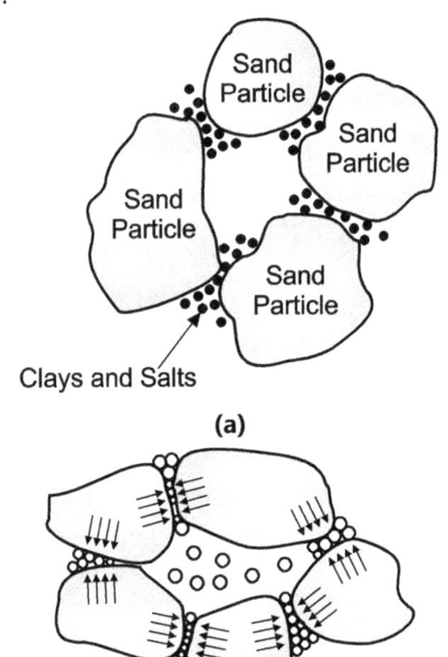

(a)

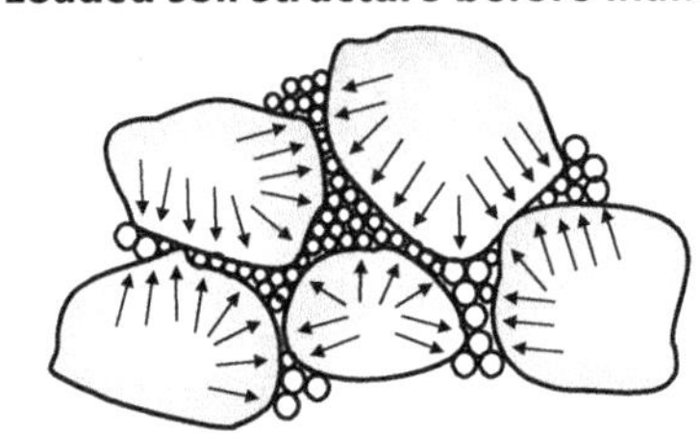

(b) : Loaded soil structure before inundation

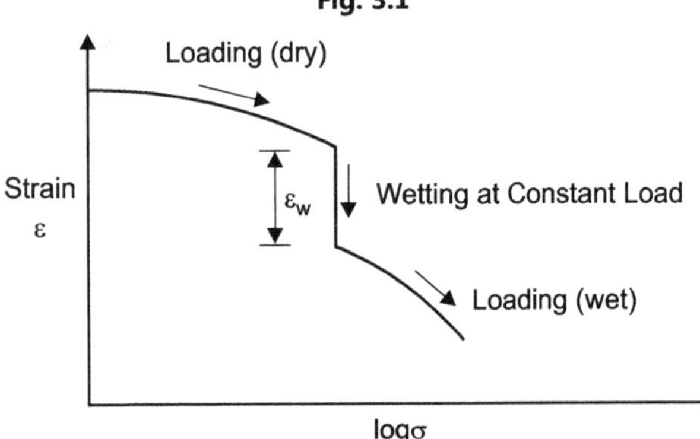

(c) : Loaded structure after inundation

Fig. 3.1

Fig. 3.2 : Single oedometer tests

Table 3.1 : Classification of Soil Collapsibility

Potential Hydro Collapse Strain, $ε_n$	Severity of Problem
0 – 0.01	No problem
0.01 – 0.05	Moderate trouble
0.05 – 0.10	Trouble
0.10 – 0.20	Severe trouble
> 0.20	Very severe trouble

3.3.1 Guidelines for Construction in Collapsible Soil

- **Wetting Depths ≤ 2 Meters**
  - ➢ Removal of the collapsible soil.
  - ➢ Avoidance or minimization of wetting.
  - ➢ Injection of chemical stabilizers or grout.
  - ➢ Prewetting.
  - ➢ Compaction with rollers or vehicles.
  - ➢ Compaction with displacement piles.
  - ➢ Controlled wetting.
  - ➢ Design structure to be tolerant of differential settlements.

- **Wetting Depth > 2m**
  - ➢ Compaction by heavy tamping.
  - ➢ Vibroflotation.
  - ➢ Deep blasting combined with prewetting.
  - ➢ Transfer of load through the collapsible soils to the stable soils below.

Fig. 3.3 : Deep compaction

Fig. 3.4 : Stone column

Fig. 3.5 : Grouting

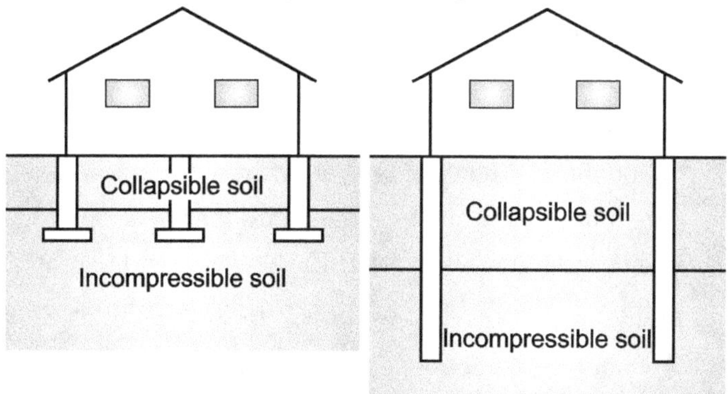

Fig. 3.6

3.4 EXPANSIVE SOIL

- Expansive soils are soils that expand when water is added, and shrink when they dry out. This continuous change in soil volume can cause homes built on this soil to move unevenly and crack.

- Expansive soils cause damage to houses, other buildings, roads, pipelines, and other structures. This damage is more than twice the damage from floods, hurricanes, tornadoes, and earthquakes combined.

- Expansive soil in India is commonly called as Black cotton soil.

- These soils cover vast areas of Maharashtra Karnataka, Andhra, Madhya Pradesh and Gujarat.

3.4.1 Causes of Moisture Changes in Soils

- Moisture is the indication of amount of water present in a soil mass and is usually expressed in terms of percentage on mass basis. Expansive soils contain minerals (Montmorillonite) that are capable of absorbing water. When they absorb water, they increase in volume. Various causes of change in moisture (increase / decrease) of soil are given below :

1. **Natural Causes :** Natural causes to change moisture in the soil are rain, change in temperature, poor drainage, leakage through conveyance system etc.

2. **Artificial Causes :** Artificial causes to change in moisture are construction activity can exacerbate the effect of expansive soils. For example, artificial irrigation causes more water to infiltrate the ground, while at the same time less water evaporates due to there being more roadways, parking lots, driveways, sidewalks, and buildings. This results in an increase in subsurface moisture.

3.4.2 Effects of Swelling on Buildings

- Due to increase in the moisture of soil volume of soil will increases. Expansions of ten percent or more are not uncommon. This change in volume can exert enough force on a building or other structure to cause upheaval of buildings and thus by damage it.

- Cracked foundations, floors, and basement walls are typical types of damage done by swelling soils. Damage to the upper floors of the building can occur when motion in the structure is significant.

- Expansive soils will also shrink when they dry out. This shrinkage can remove support from buildings or other structures and result in damaging subsidence.

- Fissures in the soil can also develop. These fissures can facilitate the deep penetration of water when moist conditions or runoff occurs.

- This cycle of shrinkage and swelling places repetitive stress on structures, and damage worsens over time.

3.4.3 Problems Associated with Expansive Soil

- Low bearing capacity.
- High compressibility (CH-MH).
- Swelling and swelling pressure.
- Shrinkage.

3.4.4 Common Damage in Building on Expansive Soil

- Separation of wall and roof slab.
- Uplift of interior footing.
- Diagonal cracks near opening.
- Separation of interior flooring from exterior walls.
- Horizontal cracks along the longer span of slab.
- Vertical cracks between column and walls etc.

Fig. 3.7 : Expansive soil

3.4.5 Guidelines for Design / Construction in Expansive Soils

- **Remedial Measures :**
  - ➤ Remove the expansive soil.
  - ➤ Mix the soil with non-expansive material.
  - ➤ Mix in chemicals to change the way the clay reacts with water.
  - ➤ Keep the soil moisture constant.
  - ➤ Use reinforced foundations that are designed to withstand soil volume changes.
  - ➤ Ensure proper drainage away from the foundation.
  - ➤ Provide under-reamed pile.

- **Preventive Measures :**

There are many ways to prevent damage to structures from expansive soils. These methods can be classified as follows:

1. Controlling water's access to the soil;
2. Altering the soil properties;
3. Altering the method of construction

1.   Controlling Water's Access to the Soil :

The primary idea here is to keep all water away from the soil surrounding the foundations. This involves such common sense measures as:

- Controlling surface drainage by making sure that the ground surface slopes away from the structure and the foundation.
- Draining rainfall sufficiently far away from the foundations.
- Not placing landscaping that will require extensive watering near a structure.
- Using impervious liners that serve as moisture barriers to isolate foundations from surface water.

2.   Altering the Soil Properties (Modification of Expansive Soil)

- Excavation and replacement of the active soil; by CNS soil (**Cohesive Non-Swelling Soil**).
- Chemical treatment of the active soil (for example lime treatment or mixing the soil with PFA [Pulverized Fuel Ash]) to reduce its expansive potential; and
- Prewetting of the soil. The idea is to build the foundation on a soil that has already expanded rather than a soil that will expand after the foundation is in place.

3.   Altering the Method of Construction

- Bypassing the expansive clay soils in the active zone by resorting to deep foundations.
- When using mat or spread footings, using "waffled" foundations. The "waffle" pattern has the advantages that it allows the soil to expand into the voids, and that the foundations are quite stiff. This reduces differential settlement and heave due to expansion and shrinkage.
- Anchoring the foundation at depth of little volume change (more than 2m).
- Construct highly flexible structures and foundations which can accommodate large settlement/movement
- Counteracting swelling pressure.
- Stiffening of structural components by providing bands at various levels.

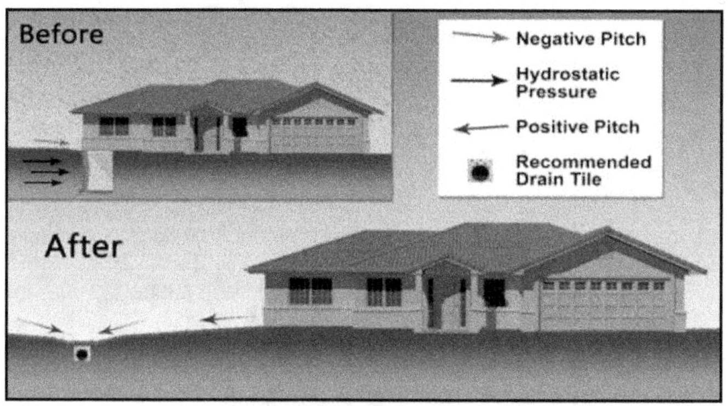

Fig. 3.8

Fig. 3.9

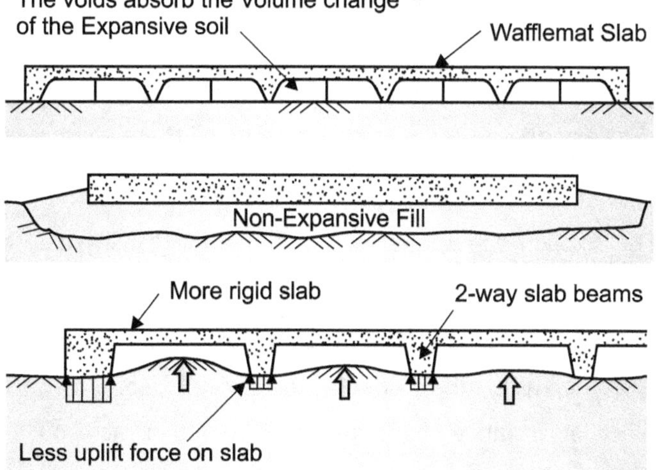

Fig. 3.10 : Soil expands into voids

3.5 CORROSIVE SOIL

Schiff (1982) indicated that corrosion would be most likely in the following soil conditions.

- High moisture content
- Poorly aerated
- Fine grained
- Black or grey colour
- Low electrical resistivity
- Low or negative redox potential
- Organic material present
- High chemical content
- Highly acidic
- Sulfides present
- Anaerobic microorganism present

Corrosion is the problem for structure where steel is used steel foundations in marine environment have a significant potential for corrosion especially those exposed to salt water. Areas where the elevation of the groundwater table fluctuates, such as tidal zones are potential areas for corrosion. Contaminated soils, such as sanitary landfills and shorelines near old outfalls, are also more likely to have corrosion problems.

3.5.1 Preventive Measures for Corrosion

- **Steel Structures:**
  - ➢ Use different construction material (concrete or wood).
  - ➢ Increase the thickness of steel section by an amount equal to anticipated deterioration.
  - ➢ Cover steel with protective coating (coal tar epoxy).
  - ➢ Provide cathodic protection.
- **Concrete Structures:**
  - ➢ Buried concrete is usually very resistant to corrosion and will remain intact for many years. However, presence of sulphates (SO_4) in soil can react with the cement to form calcium sulfo-aluminate (ettringite). As these crystals grow and expand, the concrete cracks and disintegrates.
  - ➢ Reduce the water cement ratio (reduce permeability).
  - ➢ Increase cement content (reduce permeability).
  - ➢ Use sulphate resisting cement.
  - ➢ Coat the concrete with an asphalt emulsion.

3.6 GROUND IMPROVEMENT METHODS

- Ground improvement are the techniques where in unsuitable soil is made suitable by changing its properties (physical / chemical) In general terms ground improvement may be considered to be "When the Engineer forces the soil to adapt to the project requirements by altering its natural state, rather than changing the Engineering design in response to the natural limitation of the soil."

Table 3.2 : Techniques of Ground Improvement

Reference	Criteria	Categories
Mitchell (1981)	Construction / function	• In situ deep compaction of cohesion less soils • Pre-compression • Injection and grouting • Admixtures • Thermal • Reinforcement
Hausmann (1990)	Process	• Mechanical • Hydraulic • Physical and chemical • Modification by inclusion and confinement
Ye et al. (1994)	Function	• Replacement • Deep densification • Drainage and consolidation • Reinforcement • Thermal treatment • Chemical stabilization
Chu et al. (2009)	Soil type and inclusion	• Ground improvement without admixtures in granular soil or fill material • Ground improvement without admixtures in cohesive soils • Ground improvement with admixtures or inclusions • Ground improvement with grouting type admixtures • Earth reinforcement
Schaefer and Berg (2012)	Application	• Earthwork construction • Densification of cohesion less soil • Embankment over soft soils • Cutoff walls • Sustainability • Soft ground drainage and consolidation • Construction of vertical support elements • Lateral earth support • Liquefaction mitigation • Void filling
Jie Han (2015)	Function	• Densification • Replacement • Drainage and consolidation • Chemical stabilization • Reinforcement • Thermal and biological treatment

EXERCISE

1. What is an expansive soil?

2. What are the characteristics of expansive soil?

3. What are the problems associated with expansive soil?

4. What is difficult soil? What are various difficult soil any one problem associated with all these soils

5. What is the mineral present in expansive soil?

6. Explain the spatial distribution of expansive soil in India

7. What are the construction techniques that will minimize the effect of difficult soil on the performance of structure constructed in such soil?

8. List typical damage that a structure will face when it is constructed in expansive soil

9. What is CNS ?

10. Can we reduce the damage to structure resting on difficult soil by using geo-synthetics.

◈ ◈ ◈

SHALLOW FOUNDATION

4.1 INTRODUCTION

The foundation in which depth of foundation is less than or equal to width of foundation is known as 'shallow foundation'. It is commonly used for structure carrying moderate loads and subsoil is having sufficient strength.

Types of Shallow Foundation

- The foundation in which depth of foundation is less than or equal to width of footing is called shallow foundation which is further subdivided into following types:

  ➤ Isolated footing.

  ➤ Combined footing.

  ➤ Strap footing.

  ➤ Strip or continuous footing (wall footing).

  ➤ Mat or raft footing.

- Footing which is used for supporting single column is called **'Isolated Footing'** and can be provided in any of following way:

  ➤ Spread footing,

  ➤ Stepped footing or

  ➤ Slopped footing.

- Footing which supports two adjacent columns is called **'Combined Footing'** and is generally rectangular or trapezoidal.

- Footing in which two isolated footings are connected by a structural member (strap beam) is called **'Strap Footing'**.

- Footing which is used for supporting wall or row of columns is called **'Strip or Continuous Footing'**.

- Footing which supports walls and columns under part or all of the structure, is termed as **'Mat or Raft Footing'**.

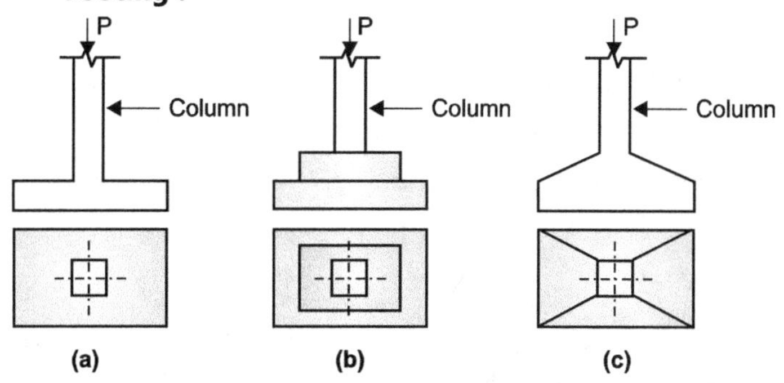

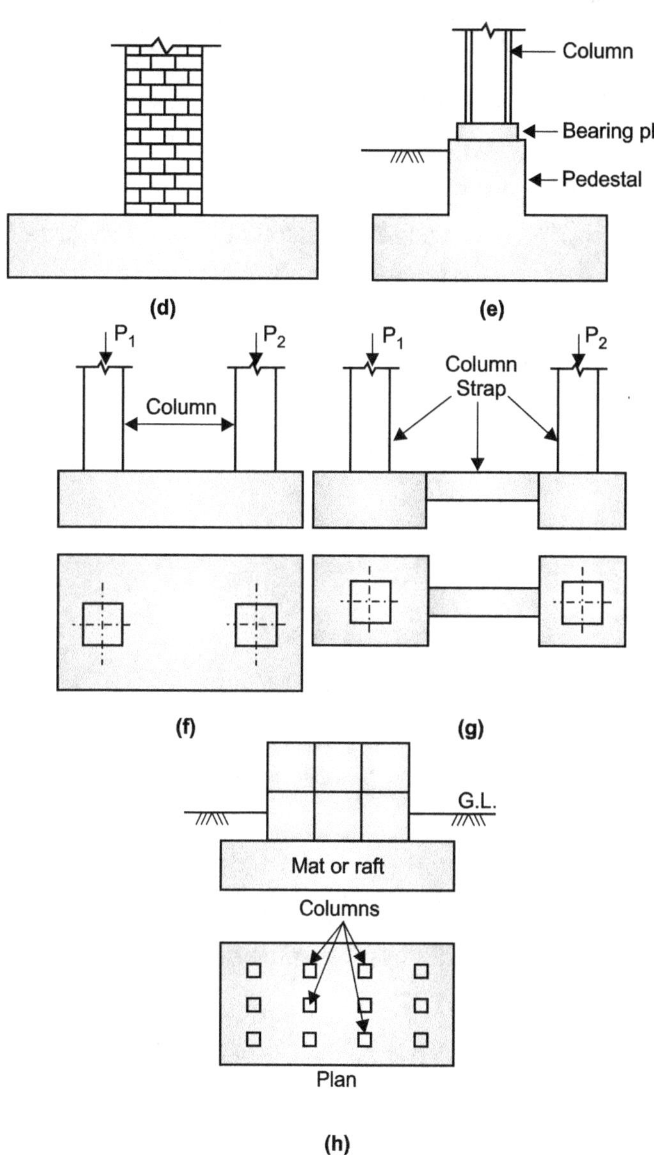

Fig. 4.1 : Common type of footings : (a) single spread footing, (b) stepped footing,(c) Sloped footing, (d) Wall footing, (e) Footing with pedestal, (f) Combined rectangular footing, (g) Strap footing, (h) Mat or raft foundation

4.1.1 Design of Shallow Foundation

- Design of foundation is the process wherein we have to determine

  ➤ Size of footing

  ➤ Shape of footing

  ➤ Depth of footing

  ➤ Reinforcement required to resist shear and bending induced in the footing

- Determination of first three parameters is called **geometrical design** of footing and determination of reinforcement required is called **structural design** of footing.

4.2 DESIGN OF ISOLATED FOOTING

1. **Size of Footing:** Area required for a footing is calculated by using following considerations :

- **Shear Failure Consideration (Bearing Capacity):**

Area = Load on the column + [self weight of footing × safe bearing capacity of footing]

Self-weight of footing is taken as 10% of load on the column.

- **Settlement Consideration: (Housel's Method)**

Housel has given an equation to calculate load carrying capacity of footing

i.e. $Q = [(A \times m) + (P \times n)]$

where

A : Area of footing

P : Perimeter of footing,

m & n : Constants which are determined by using plate load test for two different size of plates

Let P : load for which footing is to be designed,

S : Permissible settlement for footing and soil system,

Q1 : Load on plate 1 of area A1 and perimeter P1 for a settlement of S

Q2 : Load on plate 2 of area A2 and perimeter P2 for a settlement of S

By substituting these values in Housel's equation we get following equations

$$Q1 = mA1 + nP1 \qquad ... (4.1)$$

$$Q2 = mA2 + nP2 \qquad ... (4.2)$$

Solving these two equations we get values of constants m and n

Substituting these values of m and n again in Housel's equation [for example for square footing]

$$Q = B^2 \times m + 4B \times n$$

Solving above equation we get size of footing B.

2. **Shape of Footing**

Usually square footing is preferred when load is axial or eccentric about both axes any one axis When load is eccentric about then rectangular footing is preferred

3. **Depth of Footing:**

Vertical distance between the ground surface and base of footing is called '**Depth of Footing**'.

Depth of footing depends on number of factors. Thus, for particular type of structure it is taken as maximum amongst the following :

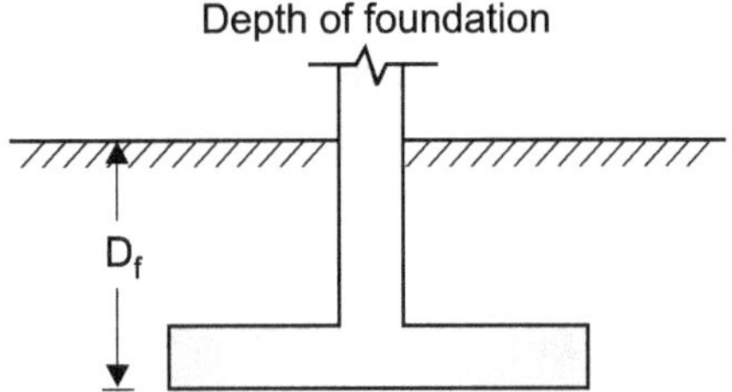

Fig. 4.2 : Depth of foundation

- **Loading Criteria:**

$$D = \frac{q}{r}\left[\frac{1 - \sin\varphi}{1 + \sin\varphi}\right]^2$$

where

q – SBC of soil,

r – unit weight of soil and

φ – angle of internal friction of soil

- **Depth of Volume Change**: (Expansive soil)

D > 1.5m,

D > Root depth,

D > Level of artificially increased temperature

- **Frost Depth:**

D > frost depth

D > 1.5m in cold areas,

- **Erosion Depth:**

D > Erosion depth,

D > 0.3m single/double storey,

D > 0.6m for heavy construction

- **Scour Depth:**

D > Scour depth

- **Settlement Consideration:**

D > Depth of top organic soil (compressible soil).

If depth of organic soil is considerable then one of following alternative can be used

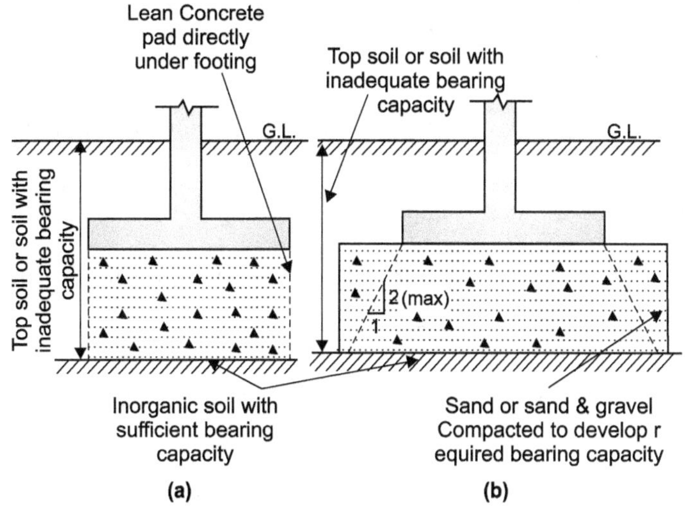

Fig. 4.3

- **Topographical Consideration:**

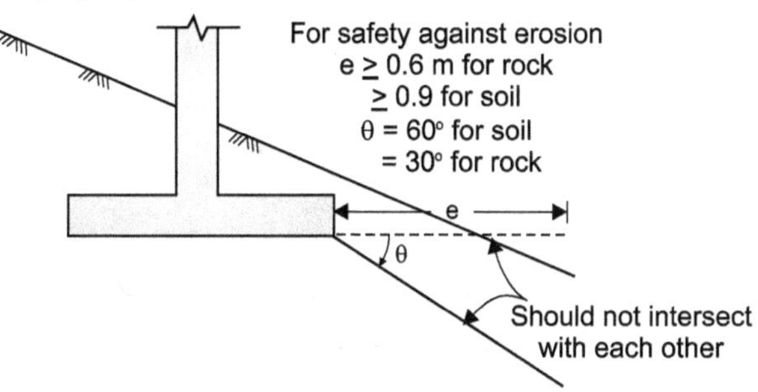

Fig. 4.4

- **Adjoining Structure:**

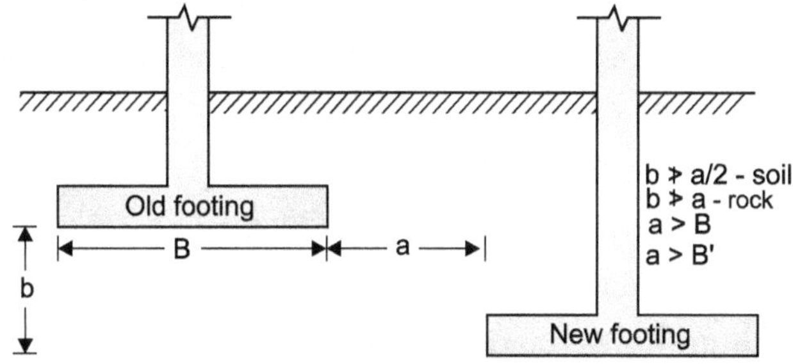

Fig. 4.5

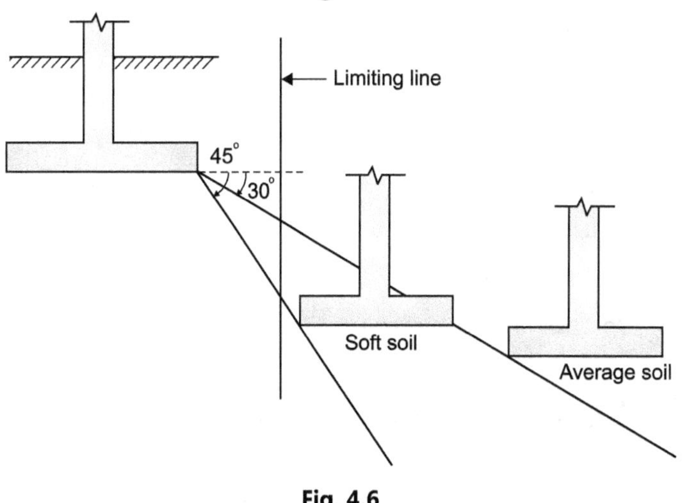

Fig. 4.6

As far as possible

- ➢ New footing should not extend beyond limiting line.
- ➢ New footing should be carried below old footing to minimize effect of new on old footing.
- ➢ New footing should not intersect reference line (red line).

- **Sloping Ground and Adjoining Structure:**

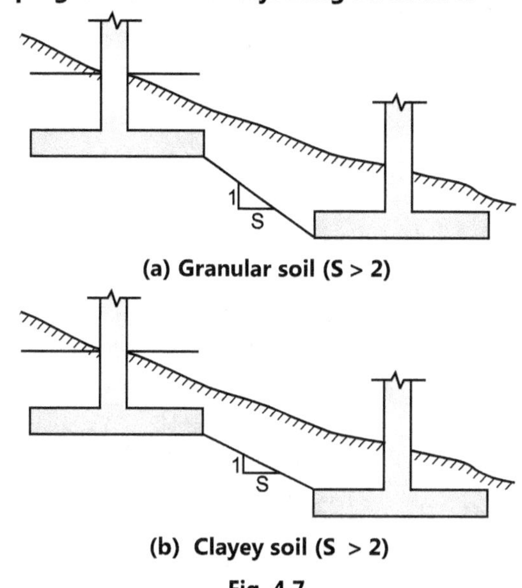

Fig. 4.7

- **Position of Ground Water Table:** Ground water table affects in three ways:

  1. Construction below Ground water table is difficult.
  2. It reduces bearing capacity of soil.
  3. Uplift and damp proofing of structure.

Note : D_f < Depth of ground water table

4.3 DESIGN OF COMBINED FOOTING

- The use of combined footings is appropriate either when two columns are spaced so closely that individual footing are not practicable or when a wall column is so close to the property line that it is impossible to center an individual footing under the column.

- Basic principle of design of foundation is that line of action of resultant load should coincide with the geometrical center of footing, so that the pressure distribution is uniform.

- A combined footing is so proportioned that the centroid of the area in contact with the soil lies on the line of action of the resultant of the loads applied to the footing; consequently, the distribution of soil pressure is reasonably uniform. In addition, the dimensions of the footing are chosen such that the allowable soil pressure is not exceeded. When these criteria are satisfied, the footing should neither settle nor rotate excessively. A combined footing may be of rectangular shape or of trapezoidal shape in plan. These are usually constructed using reinforced concrete.

- If area required for a column footing is not overlapping with area zone of adjacent footing or not crossing the boundary of plot, in such case one can use isolated column footing. However, if any one of above criteria is not satisfied then either, we have to use combined footing or strap footing. Combined footing is used when adjacent columns are close to each other, so that area of individual column footing overlaps or are very close. In order to match load center and geometrical center of footing either provide rectangular or trapezoidal combined footing.

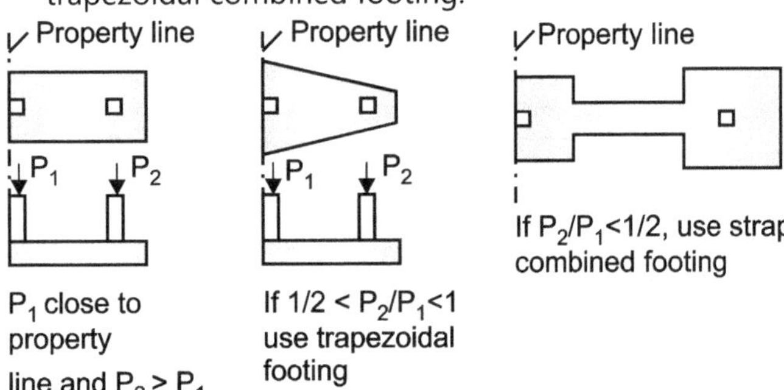

Fig. 4.8

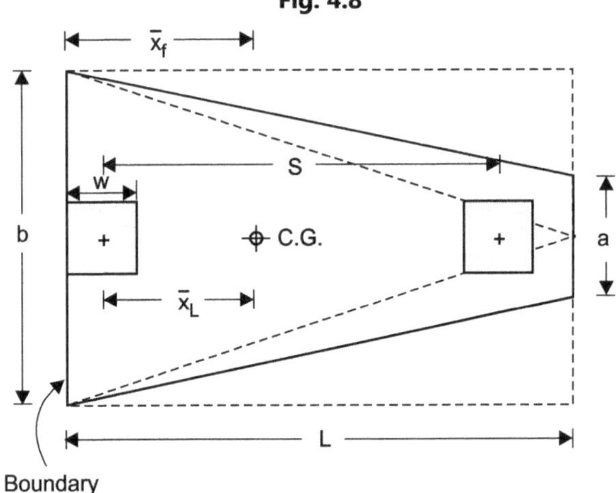

Fig. 4.9

Consider two adjacent columns having center to center spacing S, width of exterior column is w.

Let, $\bar{x}_L$ – Distance of resultant of column load from center of external column

     $\bar{x}_f$ – Distance of centroid of footing

   a and b – Sides of footing

      L – Length of footing

$$\bar{x}_f = \bar{x}_L + \frac{w}{2} \quad \text{... (From load consideration)}$$

$$\bar{x}_f = \frac{2a + b}{a + b} \cdot \frac{L}{3} \quad \text{and}$$

$$A_f = \frac{a + b}{2} \cdot L \text{... (From geometry of footing)}$$

- If $a = b$, then footing will be rectangular and $\bar{x}_f = \frac{L}{2}$.

- If $a = 0$, then footing will be triangular (practically not possible) and $\bar{x}_f = \frac{L}{3}$.

Thus, geometrical criteria's for selecting appropriate type of footing are as mentioned below.

- If $\bar{x}_f$ as calculated from load consideration is more than or equal to $\frac{L}{2}$, then provide rectangular footing.

- If $\bar{x}_f < \frac{L}{3}$, then provide strap footing.

- If $\bar{x}_f$ lies between $\frac{L}{3}$ and $\frac{L}{2}$, then provide trapezoidal footing.

Design of footing (foundation engineering) means to fix the geometrical parameter of footing in plan and to draw its shear force and bending moment diagram.

4.3.1 Rectangular Combined Footing

Footing can be rectangular under following situations:

- Two adjacent column carries equal load and the area of footing coincides.

- Restriction over width of footing.

- If interior column carries larger load than external column, which is close to boundary.

Procedure for Rectangular Combined Footing :

- **Given data :**

    1.   Allowable soil pressure.

    2.   Loads carried by column.

    3.   Distance of external column from boundary.

- **Required :**

1. To fix dimension of footing.

- **Procedure/Steps :**

1. Determine resultant column load,

$$R = Q_1 + Q_2$$

2. Determine area of footing required,

$$A_f = \frac{R}{q_a}$$

3. Locate position of resultant load with respect to external column,

$$x_L = \frac{Q_2 \cdot L}{R}$$

4. Determine position of centre of gravity of footing and length of footing,

$$x_f = x_L + \frac{w}{2}$$

$$L = 2x_f$$

5. Determine width of footing

$$B = \frac{A_f}{L}$$

6. Determine upward soil pressure per unit width of footing,

$$p = \frac{q_a}{B}$$

7. Draw shear force and bending moment diagram along the length using column loads as point load and soil pressure as uniformly distributed load.

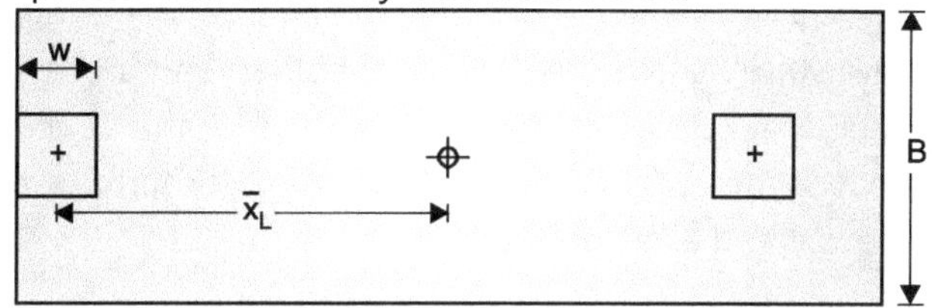

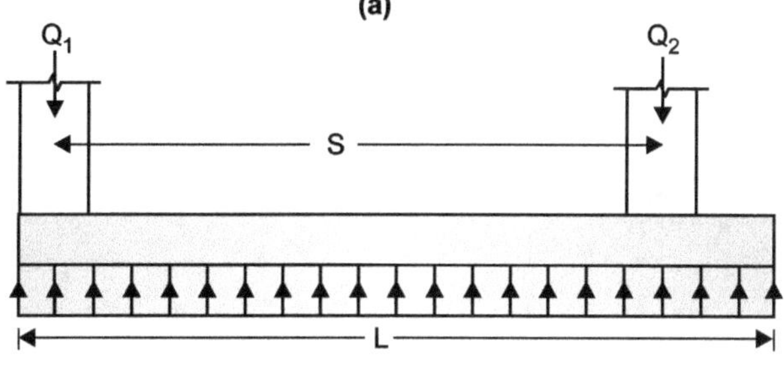

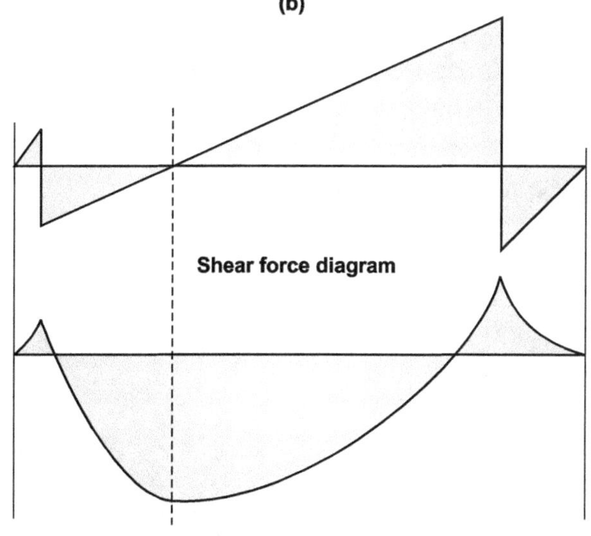

Fig. 4.10

4.3.2 Design of Trapezoidal Combined Footing (Conventional Method)

- When the two column loads are unequal, the exterior column carrying higher load and when the property line is quite close to the exterior column, a trapezoidal combined footing is used. It may be used even when the interior column carries higher load; but the width of trapezoid will be higher in the inner side.

- The location of the resultant of the column loads establishes the position of the centroid of the trapezoid. The length is usually limited by the property line at one end and adjacent construction, if any, at the other.

- The width at either end of the trapezoid can be determined from the solution of two simultaneous equations—one expressing the location of the centroid of the trapezoid and the other equating the sum of the column loads to the product of the allowable soil pressure and the area of the footing. The resulting pressure distribution is linear or uniformly varying (and not uniform) as shown in Fig. 4.11.

Procedure for Traperoidal Combined Footing :

Given Data :

1. Loads on columns,
2. Spacing between columns,
3. Allowable soil pressure and
4. Distance of external column from boundary and length of footing.

Required:

1. Size of footing,
2. Shear Force Diagram (SFD) and
3. Bending Moment Diagram (BMD).

Steps:

1. Determine resultant load, $R = Q_1 + Q_2$.

2. Find area of footing, $A_f = \dfrac{R}{q_a}$

3. Locate position of resultant load, $\bar{x}_L = \dfrac{Q_2 \cdot S}{R}$.

4. Find center of gravity (C.G.) of footing, $\bar{x}_F = \bar{x}_L + \dfrac{w}{2}$.

5. Compare $\bar{x}_f$ with $\dfrac{L}{2}$ and $\dfrac{L}{3}$, if $\dfrac{L}{3} < \bar{x}_f < \dfrac{L}{2}$ provide trapezoidal footing.

6. If a, b and L are dimensions of footing.

$$A_f = \frac{a+b}{2} \cdot L$$

and $\qquad \bar{x}_f = \dfrac{2a + b}{a + b} \dfrac{L}{3}$

Substitute value of A_f and $\bar{x}_f$ from step (2) and (4) and solve above equation for a and b.

7. Compute the intensity of upward soil pressure at each edge of footing and also under load.

$$p_b = (q_a)\, b \quad \text{and} \quad p_a = (q_a)\, a$$

p_1 and p_2 can be determined by interpolation.

8. Compute shear force (S.F.) and bending moment (B.M.) at critical points and hence plot shear force and bending moment diagram.

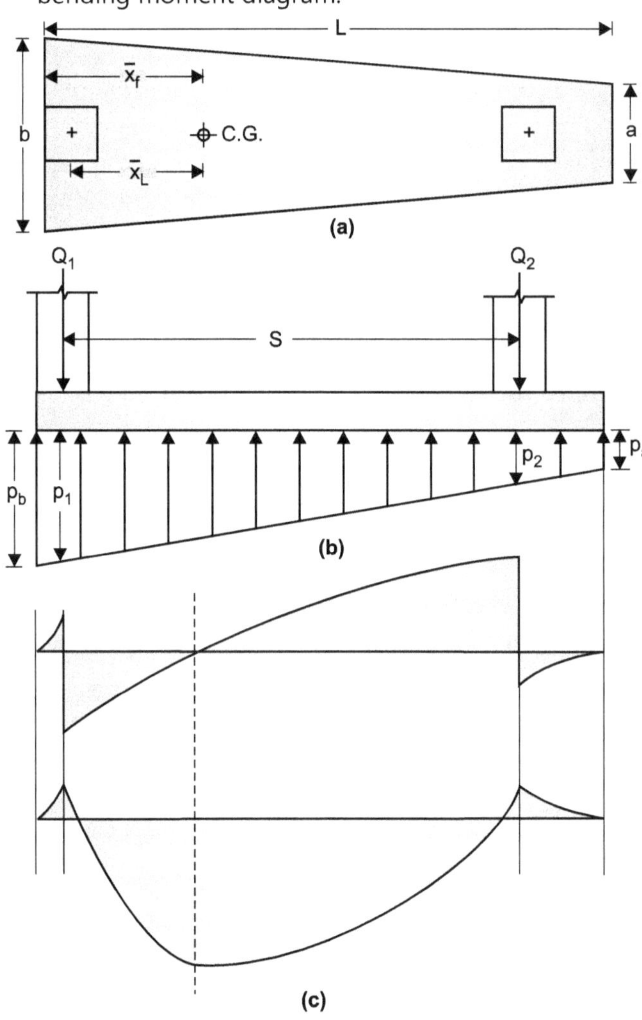

Fig. 4.11

Note: Simplified formulas for computing pressure intensity and bending moment. Due to trapezoidal loading.

$$c_x = c_2 - \frac{c_2 - c_1}{(x + y)}\,(x)$$

$$= c_1 + (c_2 - c_1)\,\frac{y}{x + y}$$

$$M_x = (2c_2 + c_x)\,\frac{x^2}{6}$$

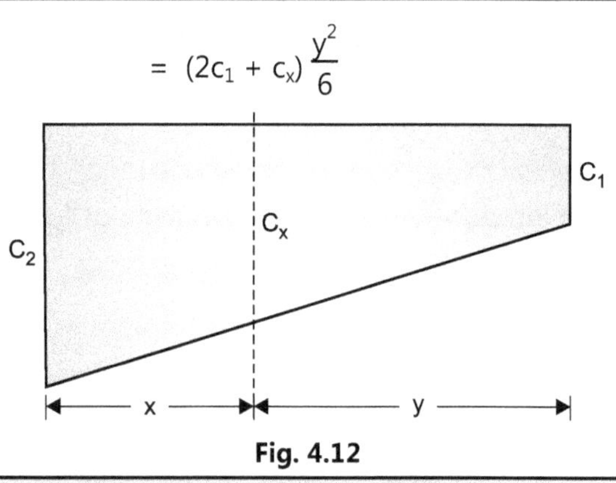

$$= (2c_1 + c_x)\,\frac{y^2}{6}$$

Fig. 4.12

4.4 DESIGN OF STRAP FOOTING

Basis for Design of Strap Footings

Strap Footings are Designed Based on the Following Assumptions:

- The strap footing is considered to be infinitely stiff. It serves to transfer the column loads onto the soil with equal and uniform soil pressure under both the footings.

- The strap is a pure flexure member and does not directly take soil reaction. The soil below the strap will be loosened up in order that the strap does not rest on the soil and exert pressure.

With these assumptions, the procedure of design is simple. With reference to Fig. it may be given as follows:

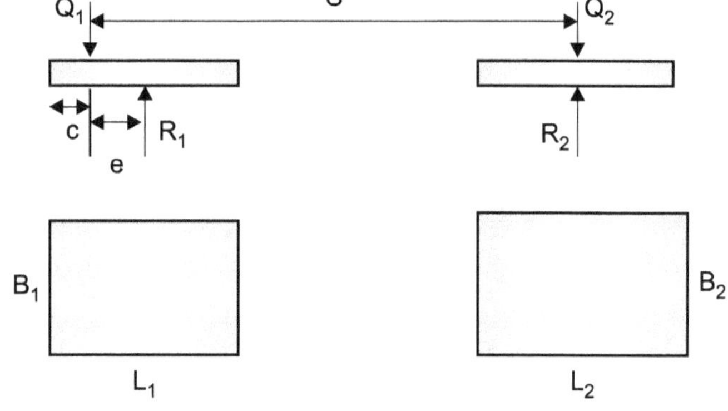

Fig. 4.13

Design Steps :

1. Assume suitable value of eccentricity e (0.5 m – 1.5 m)

2. Calculate soil reaction under each footing

$$R_1 = Q_1 \left\{ \frac{S}{S - e} \right\}$$

$$\text{and} \qquad R_2 = Q_1 + Q_2 - R_1$$

3. Calculate area of footing required to transmit soil reaction safely

$$A_1 = \frac{R_1}{SBC} \quad \text{and} \quad A_2 = \frac{R_2}{SBC}$$

4. Calculate dimensions of footing

$$L_1 = 2(c + e) ; B_1 = \frac{A_1}{L_1} ;$$

$$B_2 = B_1 \text{ and } L_2 = \frac{A_2}{B_2}$$

5. Check that L and B ratio should not be too high and also $S > 0.5 L_1 + 0.5 L_2 + e$

6. If above conditions are not satisfied then revise the design by assuming some other value of e

4.5 MAT FOUNDATION

- A mat foundation is a large concrete slab used to interface one column, or more than one column in several lines, with the base soil. It may encompass the entire foundation area or only a portion. A mat may be used to support on-grade storage tanks or several pieces of industrial equipment. Mats are commonly used beneath silo clusters, chimneys, and various tower structures.

- A mat foundation may be used where the base soil has a low bearing capacity and/or the column loads are so large that more than 50 percent of the area is covered by conventional spread footings. It is common to use mat foundations for deep basements both to spread the column loads to a more uniform pressure distribution and to provide the floor slab for the basement.

Conditions for Mat Foundations:

- Structural loads are so high and soil conditions are so poor that spread footings would be exceptionally large

- Soil is very erratic or prone to differential settlements, or soils that are expansive or subject to differential heaves.

- Structural loads are erratic.

- Unevenly distributed lateral loads (cause differential horizontal movements in spread footings or pile caps).

- Uplift loads are larger than spread footings can accommodate; weight of the mat is a factor here.

- Bottom of the structure is located below the GWT, so waterproofing is an important concern. Mat foundations are monolithic, so easier to waterproof. Weight of mat also resists hydrostatic uplift forces

Ribbed mats, consisting of stiffening beams placed below a flat slab are useful in unstable soils such as expansive, collapsible or soft materials where differential movements can be significant (exceeding 12 mm).

4.5.1 Types of Rafts

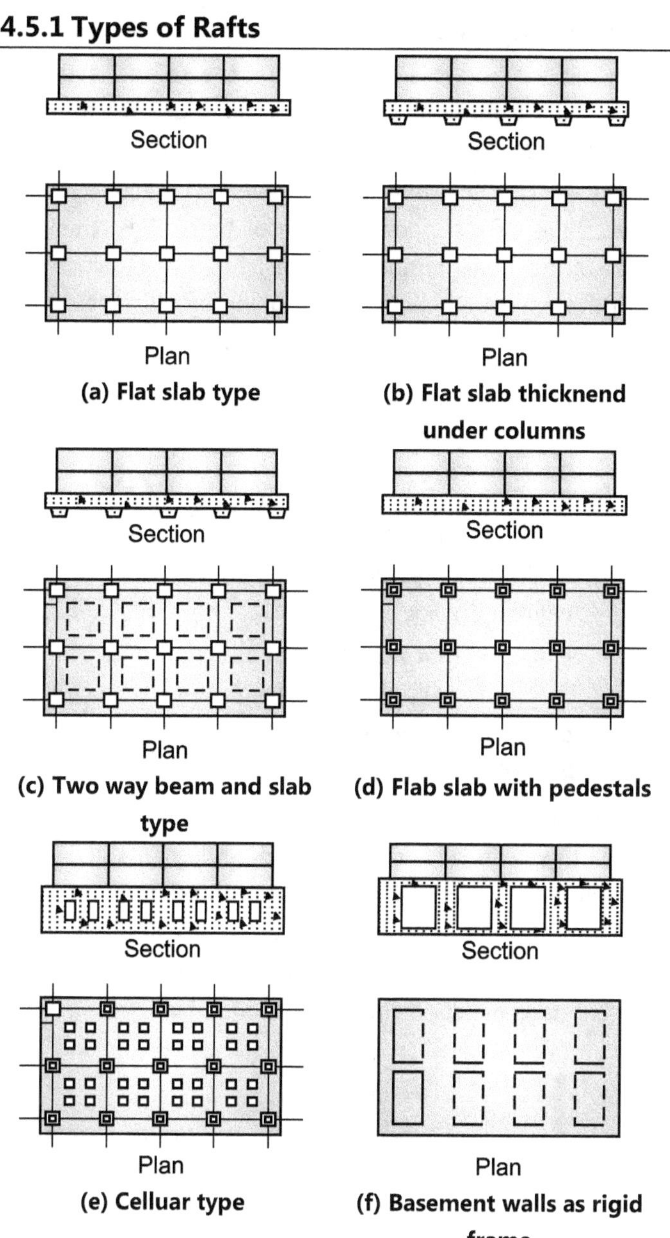

Fig. 4.14

- **Flat Slab Type :** This type of mat is used when loads on column are approximately, soil is homogeneous and columns are equispaced.

- **Slab Thickened Under Column :** This type of mat is used when columns carry heavy loads.

- **Slab and Beam Type Mat :** This type of mat is used when column carries loads with moments.

- **Cellular Mat :** This mat is used when structure carries large lateral loads along with vertical load.

4.5.2 Design of Rafts

Rafts can be designed by using following methods

1. **Elastic Method :** (Flexible method)

2. **Rigid Method :** (Conventional method/IS code method

1. Elastic Method :

- In this method it is assumed that the soil behaves like an infinite number of individual springs, not affected by adjacent springs.

- The elastic constant of the spring is equal to the modulus of subgrade reaction of the soil. Further it is assumed that spring will resist both tension and compression. This concept is given by Winkler so this method is also called Winkler approach.

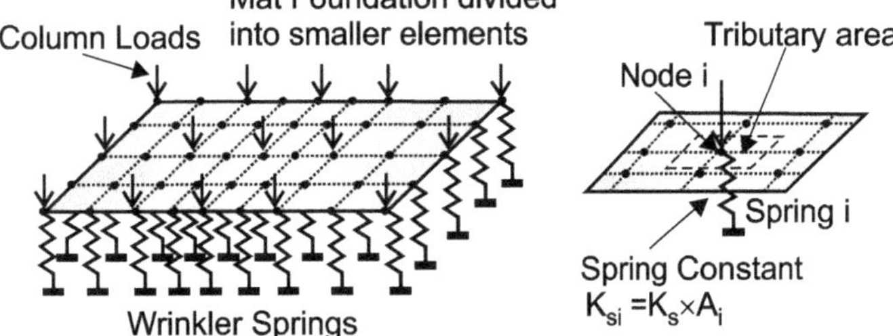

Fig. 4.15 : Mat supprted on number of piles
(Winkler approach)

2. Design of Mat Foundation by IS Code Method (Rigid Method)

Rigid Beam Method : (Conventional Method) :

This method can be used when the settlements are small. According to IS 2950 (1981), this method may be used when raft satisfies the condition $\lambda L < 1.75$ where λ is the characteristic coefficient and L is the column spacing.

The charatercistic coefficient is calculated by

$$\lambda = \left[\frac{kB}{4E_cI}\right]^{0.25}$$

where,　　k : modulus of subgrade reaction corrected for width B (kN/m^3)

　　　　　B : Width of raft

　　　　　E_c : Modulus of elasticity of concrete

　　　　　　　= $5\sqrt{f_{ck}} \times 10^6$ kPa

　　　　　I : Moment of inertia of raft in m^4

Procedure for Rigid Method :

Let us consider a raft of dimension (B × L) as shown in Fig. 4.16.

　　　P_1, P_2, P_3 ... Loads on the columns

　　　R : Resultant column load

　　　I_x : MI of raft about X axis = $BL^3/12$

　　　I_y = MI of raft about Y axis = $LB^3/12$

　　M_x = Moment about X axis

　　　　= $R\,e_y + M_x$(Internal load)

　　M_y = Moment about Y axis

　　　　= $R\,e_x + M_y$(lateral load)

e_x : eccentricity of resultant load about centroid of footing about Y axis

e_y : eccentricity of resultant load about centroid of footing about X axis

1. Choose reference axis in such a way that entire raft lies in the first co-ordinate.

2. Determine position of the resultant column load with respect these axis.

$$X' = \frac{P_1x_1 + P_2x_2 + P_3x_3 + \ldots}{R}$$

and　　$$Y' = \frac{P_1y_1 + P_2y_2 + P_3y_3 + \ldots}{R}$$

3. Determine eccentricity with respect to centre of footing.

$$e_x = X' - \frac{B}{2} \text{ and } e_y = Y' - \frac{L}{2}$$

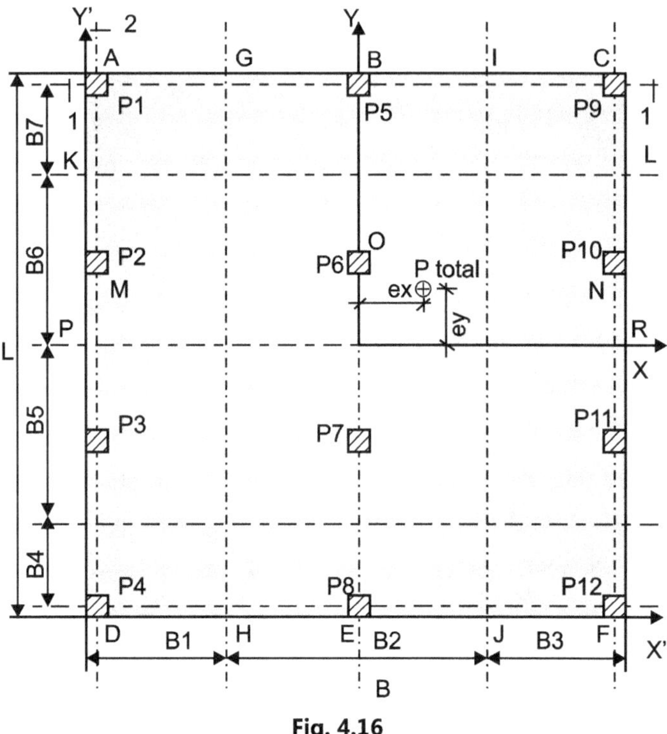

Fig. 4.16

4. Divide the raft into number of strips parallel to Y axis (B1, B2, B3) as well as parallel to X axis (B4, B5, B6 and B7) as shown in Fig. 4.16.

5. Draw SFD and BMD for all sucn strips

6. Sample calculation for one such strip GBIJEH is shown below. Soil pressure along the centre of this strip is assumed to be constant along the width of strip. Pressure at point B and E can be calculated by using the equation.

$$q = \frac{R}{A} \pm \frac{M_y}{I_y}x \pm \frac{M_x}{I_x}y$$

Calculate average pressure on the strip

$$q_{av} = \frac{q_B + q_E}{2}$$

This average pressure is used for the analysis of strip

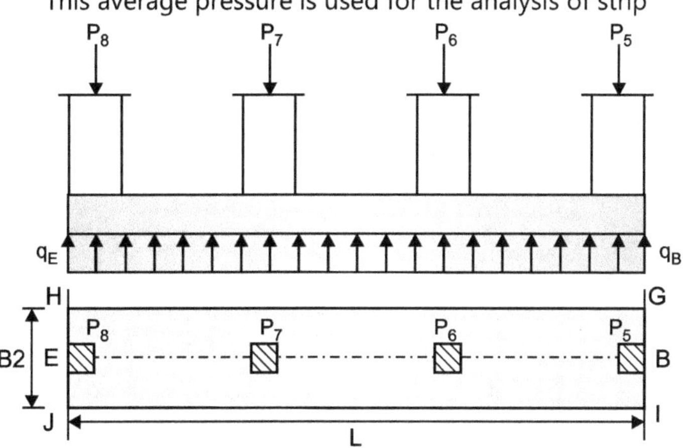

Fig. 4.17

7. The strip has to satisfy equilibrium condition

Total downward load

$$= Q_D = P_5 + P_6 + P_7 + P_8$$

Total upward load

$$= Q_U = q_{av} (L\, B_2)$$

For equilibrium $Q_D = Q_U$; However this condition is not satisfied so we need to

Average load on strip $Q_{av} = \dfrac{Q_U + Q_D}{2}$

Modified soil pressure per unit width $= w_m = \dfrac{Q_{av}}{L}$

Each column need to be multiplied by a factor α so that total upward load matches with column loads

$$\alpha\,(P_5 + P_6 + P_7 + P_8) = Q_{av}$$

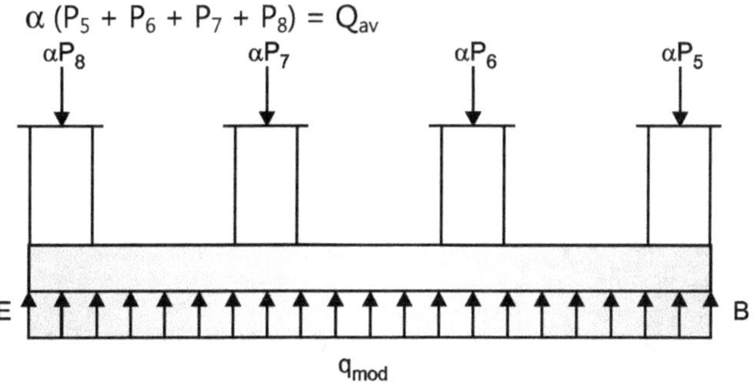

Fig. 4.18

8. Draw shear and BM diagram for the strip

Assumptions and limitation of rigid design analysis

Rigid Method is Based on Following Assumptions :

- The footing or mat is infinitely rigid, and therefore, the deflection of the footing or mat does not influence the pressure distribution. Thus magnitude and distribution of bearing pressure depends only on load.

- The soil pressure is distributed in a straight line or a plane surface such that the centroid of the soil pressure coincides with the line of action of the resultant force of all the loads acting on the foundation.

Limitations of Rigid Analysis:

- In reality no foundation is rigid some flexibility will be there which depends on thickness of foundation. In case of Mat foundation ratio of width to thickness is quite large as compared to spread footing thus assumption of rigidity does not hold good in case of raft foundation.

- Portion of mat beneath columns and bearing walls settle more than the portion with less load, which means the bearing pressure will be greater beneath the heavily loaded zones.

- Rigid method does not account for redistribution of bearing pressure (contact pressure), it does not produce reliable estimates of the shears, moments and deformation of the mat. Actual distribution of contact pressure below the footing depends on type of footing, depth of footing and type of soil

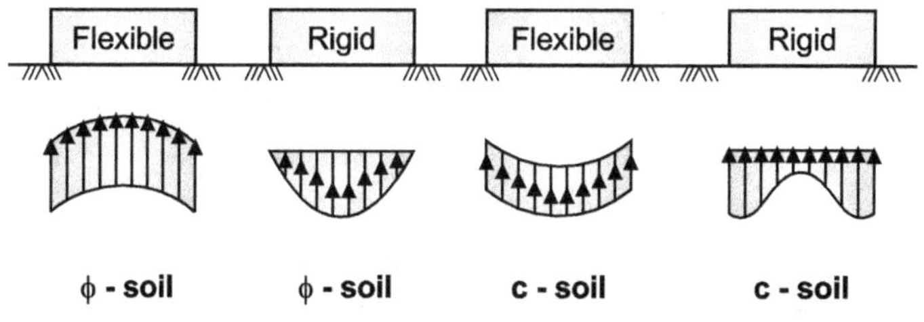

Fig. 4.19 : Contact pressure diagram for surface footing

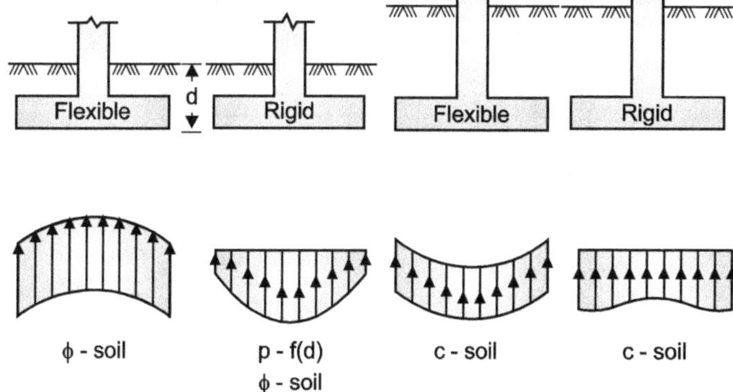

Fig. 4.20 : Contact pressure diagram for deep footing:

- In general, pressure distribution is assumed to be uniform for axial load and linearly varying for eccentric loads. The reason for such disparity between theory and practice is due to:

  ➢ Footings are neither absolutely rigid nor completely flexible.

➤ There is always certain embedment to footings. The effect of the embedment is to make soil pressure uniform.

➤ The settlement and also the bearing capacity of footing are governed by stress distribution in soil zone extending below the base of footing and contact pressure does not influence them.

- In case of flexible footing, pressure is uniform. It will produce a dish shaped settlement in clay and in case of granular soil 'E' increases with confining pressure and further E varies across width of loaded area being more near the center than near the edges, which causes dish shaped settlement.

- For a rigid footing the settlement is uniform over the contact area. Since uniform contact pressure produce dish shaped pattern thus the contact pressure must be more near the edges and less near the center for clayey soil (E = constant), and is zero at edges to maximum at center for sandy soil, because E increases towards footing center.

SOLVED EXAMPLES

Example 4.1: *Design a combined rectangular footing to carry column loads of 3000 kN and 4500 kN at 6 m spacing on a Sandy soil with allowable soil pressure of 350 kN/m². Lighter column is at a distance of 300 mm clear from the property line. Assume column size 500 mm × 500 mm.*

Solution :

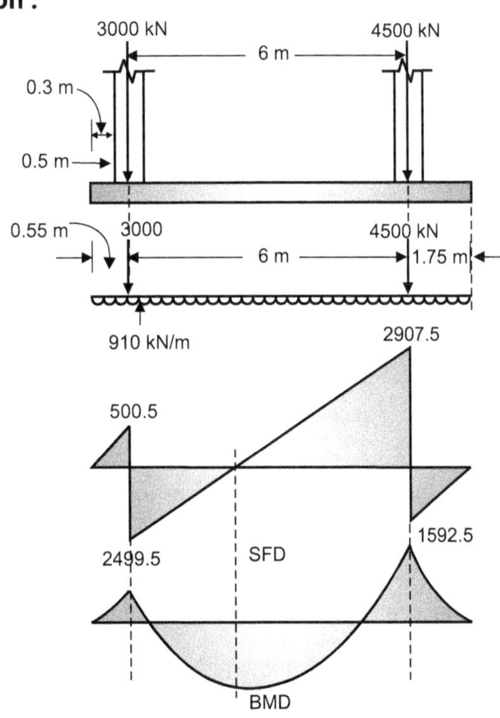

Fig. 4.21

1. Resultant load, R = 3000 + 4500 = 7500 kN

2. Position of resultant load,

$$L = \frac{4500}{7500} (6) = 3.6 \text{ m}$$

$$f = 3.6 + \frac{0.5}{2} = 0.3 = 4.15 \text{ m}$$

3. Length of footing,

$$L = 2f = 8.30 \text{ mm}$$

4. Area of footing required,

$$A_f = \frac{R}{q_a} = \frac{7500}{350} = 21.43 \text{ m}^2$$

5. Width of footing required,

$$B_f = \frac{A_f}{L} = \frac{21.43}{8.3}$$

$$= 2.58 \cong 2.60 \text{ m}$$

6. Upward soil pressure per unit lengths,

$$p = q_a \cdot B_f = 350 (2.6) = 910 \text{ kN/m}$$

S.F. and B.M. calculation,

S.F. to left of 3000 kN = 910 ×0.55 = 500.5

S.F. to right of 3000 kN = 500.5 – 3000 = – 2499.5

S.F. to right of 4500 kN = – 910 (1.75) = 1592.5

S.F. to left of 4500 kN = – 1592.5 + 4500 = 2907.5

B.M. at 3000 kN column = 910 (0.55) (0.55/2)

$$= 137.64 \text{ kN-m}$$

B.M. at 4500 kN column = 910 (1.75) (1.75/2)

$$= 1393.44 \text{ kN-m}$$

Example 4.2: *The results of two plate load tests on a given location are as follows.*

(i) diameter = 750 mm, settlement = 15 mm, ultimate load = 150 kN.

(ii) diameter = 300 mm, settlement = 15 mm, ultimate load = 50 kN.

Determine the ultimate load on a circular footing of 1.2 m diameter causing 15 mm settlement.

Solution:

Applying Housels Method for plate (0.3m width, and 0.6m width) we get

$$150 = m(0.442) + n(2.36) \qquad \qquad \dots(1)$$

$$50 = m(0.071) + n(0.94) \qquad \qquad \dots(2)$$

Solving equation 1 and 2 we get m = 92.77 and n = 46.18

Applying Housel equation for footing we get

$$Q = 92.77(1.131) + 46.18(3.77)$$
$$Q = 279 kN$$

Example 4.3 : *The results of two plate load tests for a settlement of 25.4 mm are as given below :*

Plate Diameter	Load
0.305 m	31 kN
0.610 m	65 kN

A square column foundation is to be designed to carry a load of 800 kN with an allowable settlement of 25.4 mm. Determine the size using Housel's method.

Solution : According to Housel, safe load on footing is given by

$$Q = m \cdot A + n \cdot P$$

Applying this for plate

$$31 = m(0.073) + n(0.958) \qquad \ldots \text{(i)}$$
$$65 = m(0.292) + n(1.916) \qquad \ldots \text{(ii)}$$

Solving above equations (i) and (ii) we get,

$$m = 20.55 \text{ and } n = 30.88$$

Applying for plate,

$$800 = 20.55 B^2 + 30.88 \times 4B$$
$$B^2 + 6.01 B - 38.93 = 0$$

$$\therefore \quad B = \frac{-6.01 + \sqrt{(6.01)^2 + 4 \times 38.93}}{2.0}$$

$$= 3.92 \text{ m} \approx 3.95 \text{ m}$$

∴　Size of foundation is Say 3.95 m × 3.95 m.

EXERCISE

1. Explain the conditions under which following types of combined footings are used.

   (a) Rectangular footing.

   (b) Trapezoidal footing.

   (c) Pump handle footing.

2. Explain the method of determining bearing capacity by Housel method, carrying plate load tests by using plates of different sizes.

3. In what situations would you go for combined footing? Explain any one in detail.

4. Compare rectangular combined footing with a cantilever footing.

5. What are the various types of footings and how is the depth of footing is decided ?

6. How you determine contact pressure for a footing subjected to eccentric loads ?

7. Explain the circumstances under which strap footing is used.

8. What is raft foundation? When it is preferred ?

9. What are different types of raft foundation? When each of these types are preferred ?

10. With a neat illustrative sketch explain design of rectangular combined footing.

11. With a neat illustrative sketch explain design of trapezoidal combined footing.

12. With a neat illustrative sketch explain design of strap footing.

13. With a neat illustrative sketch explain design of raft footing.

14. What are the assumptions made in design of footing by rigid method ?

15. How do you calculate the bearing capacity of raft ?

PROBLEMS FOR PRACTICE

1. Two plate load test with square plates were conducted on a soil deposit for a 30 mm settlement following loads were obtained.

Width of Plate (mm)	Load (kN)
300	38.2
600	118.5

   Determine width of square footing which would carry a load of 1500 kN for a limiting settlement of 30 mm.

   (**Ans.** 2.66 m ≅ 2.7 m)

2. Design a footing to support two columns at c/c spacing of 6m carrying load of 3000kN and 4500kN. Lighter column is near the boundary (0.3m from edge of column) assume column size 0.5m × 0.5m and SBC = 350kPa.

   (**Ans.** Rectangular footing 2.6m × 8.3m)

3. Design a strap footing for the column shown in figure. If SBC of soil is 100kPa

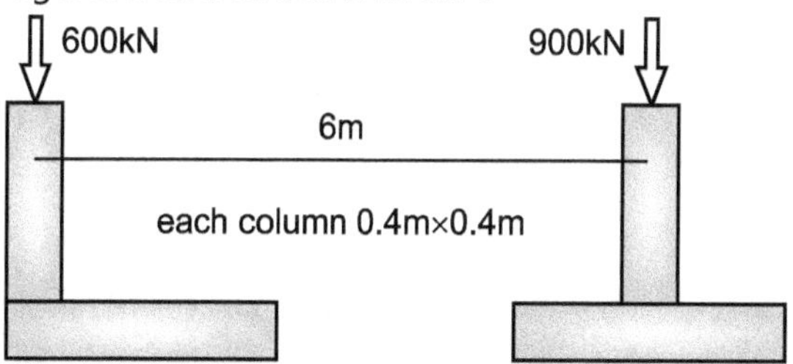

Fig. 4.22

(**Ans.** [2.6m × 3m and 2.75m × 2.75m)

4. Design combined trapezoidal footing to support two columns carrying loads of 1250kN and 950kN respectively, if the spacing c/c between columns is 3.5m size of each column is 0.3mX0.3m and columns are flushed with the footing boundary on either side

   (**Ans.** a = 4.1m, b = 1.7m and L = 3.8m)

5. Mat foundation for a structure along with the column loads is as shown in figure. Spacing c/c of columns along x axis is 6m and that along y axis is 4m edge distance in both direction is 1m. Analyze this mat and draw SFD and BMD for the strip given below. SBC of soil is 60kPa.

   (a) Extreme left strip parallel to y axis

   (b) Second strip from bottom parallel to x axis

   (c) Middle strip parallel to y axis

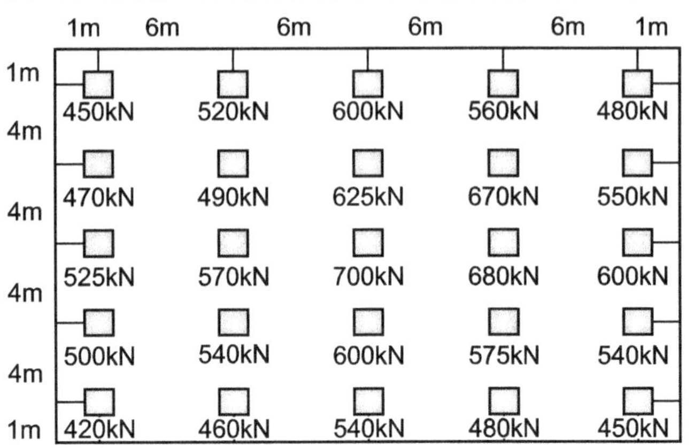

Fig. 4.23

❖ ❖ ❖

DEEP FOUNDATION

5.1 INTRODUCTION

A pile is a relatively slender member made of timber, concrete, steel or composite material usually either driven into the soil or placed/bored into the soil to provide vertical or lateral support or for other engineering functions. The piles transfer load by bearing on competent material or through the friction between the soil and the pile (skin friction).

When to use Pile Foundation?

- In general pile foundation is likely to be more expensive than spread or mat foundation. Hence are used when no alternative is left to support the structural loads; Pile foundations are used when the underlying soils are incapable of resisting the loads from the structure.

- The piling is placed in the ground through poor quality materials to bear on competent soils and is commonly used in following situations.

  ➢ Structure carries large loads and a good stratum is available at large depth.

  ➢ Structure carries large lateral loads.

  ➢ Structure carries eccentric loads.

  ➢ Structure carries inclined loads.

  ➢ Structure is highly sensitive to settlement.

  ➢ Structure is to be constructed in expansive soil with high swelling potential.

  ➢ For offshore structures.

  ➢ Bridge piers and abutments when there is large scouring.

  ➢ To anchor down the structures subjected to uplift due to water pressure or excessive moments (tension piles or uplift piles).

  ➢ For compacting loose sandy soil.

  ➢ To provide anchorage against horizontal forces from sheet piling walls or other pulling forces (anchor piles).

  ➢ To shield water front structures against impact from ships or other floating objects (fender piles and dolphins).

5.1.1 Classification of Pile

Pile foundations can be categorized into two general types:

1. Displacement piles and
2. Replacement piles.

- A displacement pile is a pile that is driven or vibrated into the ground and displaces the surrounding soil during installation. (Which in turn is classified into two types based on volume of soil displaced by piles i.e. small displacement and large displacement piles).

- A replacement pile is a pile that is placed or constructed within a previously drilled borehole and replaces the excavated soil.

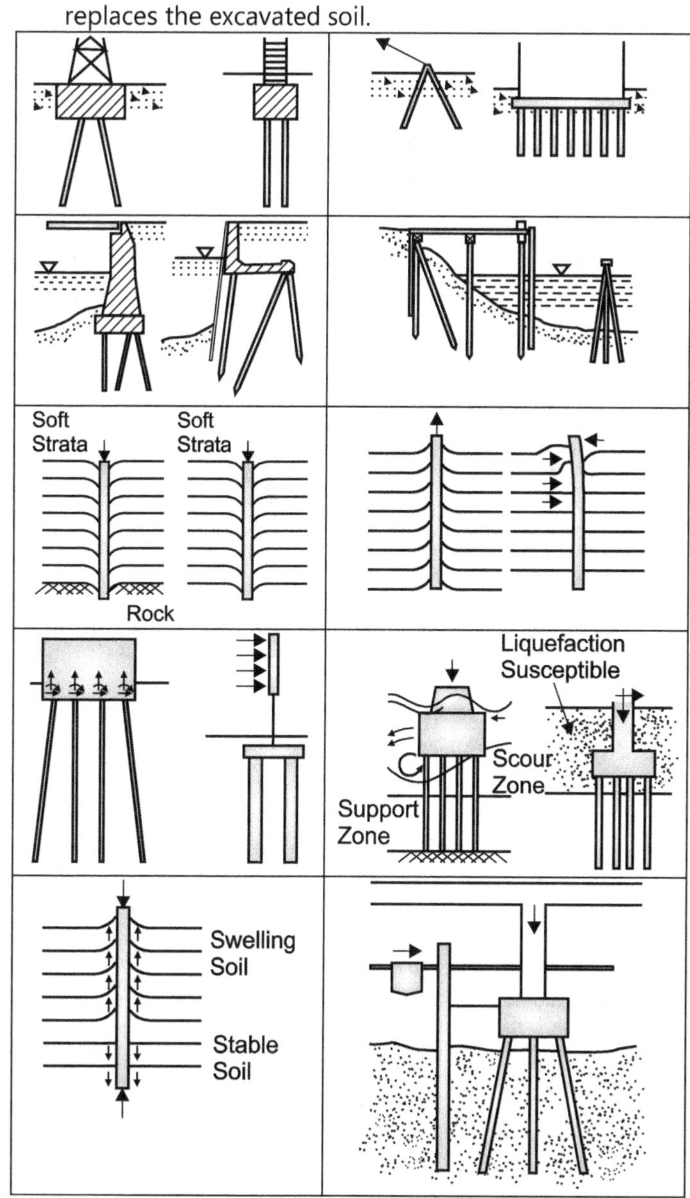

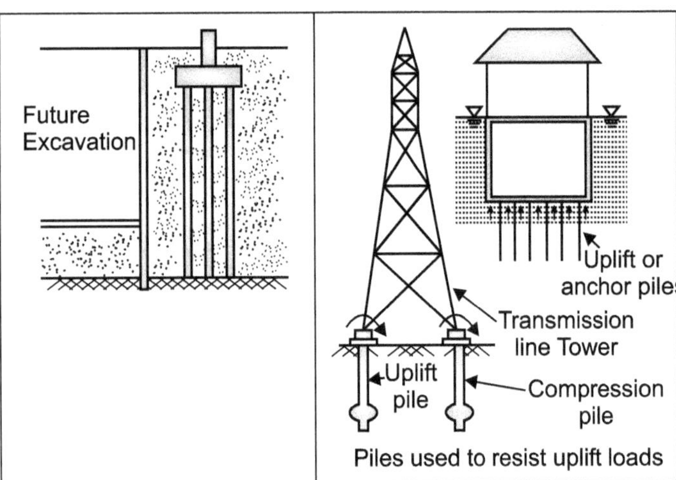

Fig. 5.1

➢ **Large Displacement** piles comprise solid-section piles or hollow-section piles with a closed end, which are driven or jacked into the ground and thus displace the soil. All types of driven and cast-in-place piles come into this category.

➢ **Small-Displacement** piles are also driven or jacked into the ground but have a relatively small cross-sectional area. They include rolled steel H- or I-sections, and pipe or box sections driven with an open end such that the soil enters the hollow section. Where these pile types plug with soil during driving they become large displacement types.

- **Replacement Piles** are formed by first removing the soil by boring using a wide range of drilling techniques. Concrete may be placed into an unlined or lined hole, or the lining may be withdrawn as the concrete is placed. Preformed elements of timber, concrete, or steel may be placed in drilled holes.

Piles Can be Classified Based on Various Criteria's as Listed Below :

Table 5.1

Classification Based on					
1. Material	Timber	Steel	Concrete		Composite
2. Shape in elevation	Cylindrical	Tapered	Under-reamed		
3. Shape in Plan	Circular	Square	Hexagon	Octagon	H-Pile	I Pile Box
	Pipe				
4. Method of Installation	Bored	Driven	Jacked	Vibrated	Jetted	Tremie

...Conti.

5. Mode of load transfer	Friction	Bearing	Combined				
6. Method of forming	Cast – in - situ	Pre–cast					
7. Length	Short	Long	Slender				
8. Volume of soil displaced	High	Low	Non				

5.2 PILE DRIVING

- In case of precast piles (concrete / steel / timber), the piles are casted and carried to the site and then is driven into the ground at desired location.

- The rig used for driving the pile is as shown below. Various kinds of rammers are used to drive the pile into ground.

- Hammers are lifted up by some distance and is then dropped on the pile, due to impact of the hammer pile will penetrate into the ground. The process is continued till pile penetrate to desired depth.

The following procedure is generally used during the driving of piles.

1. A stake is driven at the location of the pile.

2. The pile is straightened and kept upright on the location marked.

3. The plumb ness of the pile is checked.

4. The pile-driving hammer is lowered to the top of the pile.

5. Few light blows are given and checked for plumb ness.

6. Full driving starts.

7. The rate of penetration is recorded. (Rate of penetration=Number of blows per mm)

8. The pile is driven to the planned depth. In some cases, pile driving is stopped when the required rate of penetration is achieved.

9. The inspector should keep an eye on rate of penetration of the pile.

10. Unusually high blows per mm may indicate boulders or bedrock. Some piles (especially timber piles) get damaged if they hit a boulder.

11. After the pile is driven as per the required criteria, the pile is cut off from the top.

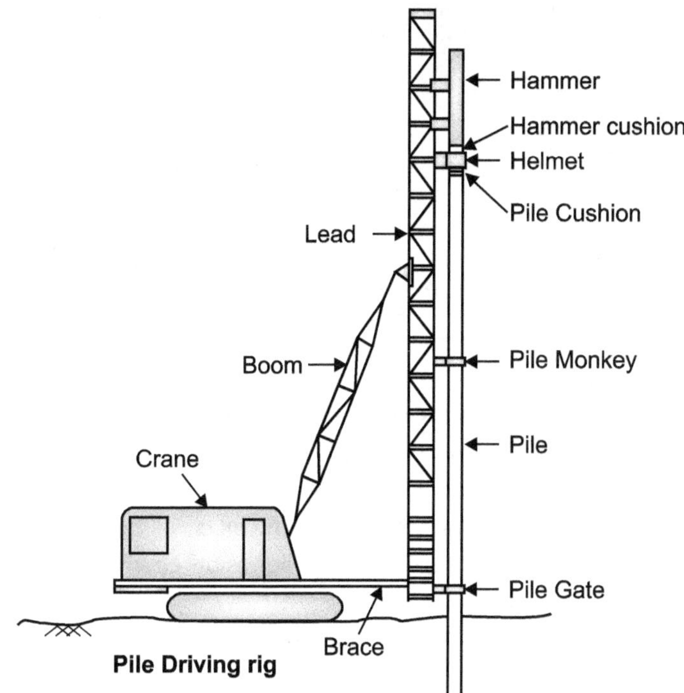

Fig. 5.2

- Important elements of the driving system include the **leads**, the **hammer cushion**, the **helmet**, and for concrete piles, the **pile cushion**. Typical components of a pile driving system are shown in Figure.

- The leads are used to align the hammer and the pile such that every hammer blow is delivered concentrically to the pile system. The helmet holds the top of the pile in proper alignment and prevents rotation of the pile during driving.

- When the hammers are used to install a pile, the following important factors must be considered:
  - ➢ Weight and size of the pile
  - ➢ Driving resistance and net transfer red energy
  - ➢ Available space at site
  - ➢ Crane facility
  - ➢ Noise control or restrictions

5.3 CAPACITY OF A PILE

Maximum load a pile can carry without failure is called its capacity and can be determined by following methods

1. Static method.
2. Dynamic method.
3. By pile load test.
4. Correlation with penetration test data

1. Static Method:

- When a load is applied at the top of a pile, the pile will tend to move vertically downward relative to the surrounding soil. This will cause shear stresses to develop between the soil and surface of the shaft.

- As a result, applied load is distributed as friction load along a certain length of the pile measured from the top.

- As the load at the top is increased, the friction load distribution will extend more and more towards the tip of the pile, till at a certain load level, the entire length of the pile is involved in generating the frictional resistance.

In general

$$Q_u = Q_p + Q_f - W_p$$

where,
Q_u : Ultimate capacity of pile
Q_U : Point Bearing load
Q_f : Friction component of load
W_p : Self weight of pile

Magnitude of W_p is very less compared to the other terms so in most of the cases it is neglected.

$$Q_u = Q_p + Q_f$$
$$Q_u = q_p A_p + f_s A_s$$

where,
$q_p = cN_c + qN_q + 0.5\gamma BN\gamma$
$f_s = \alpha c_1 + Kq_{av} \tan \delta$

Various terms in above equations are

q_p : ultimate bearing capacity of the soil at tip of pile

f_s : unit skin friction developed along the shaft of pile (< 100kPa)

N_c, N_q and N_r : bearing capacity factors

B : diameter / size of pile

K : coefficient of earth pressure

δ : angle of wall friction

q : effective overburden pressure at the tip of pile

q_{av} : average effective overburden pressure over the embedded length of pile

c : unit cohesion at the tip of pile

c_1 : average unit cohesion for the depth of embedment of pile

α : adhesion coefficient

The value of third term in the equation

$$q_p = cN_c + qN_q + 0.5\gamma BN\gamma$$

is negligible in comparison with first two terms thus the equation for point bearing can be written as

$$q_p \cong cN_c + qN_q$$

where
q : effective surcharge and is calculated as
as
q : $\gamma'L$ for $L < D_c$
and
q : $\gamma'D_c$ for $L > D_c$
D_c : critical depth

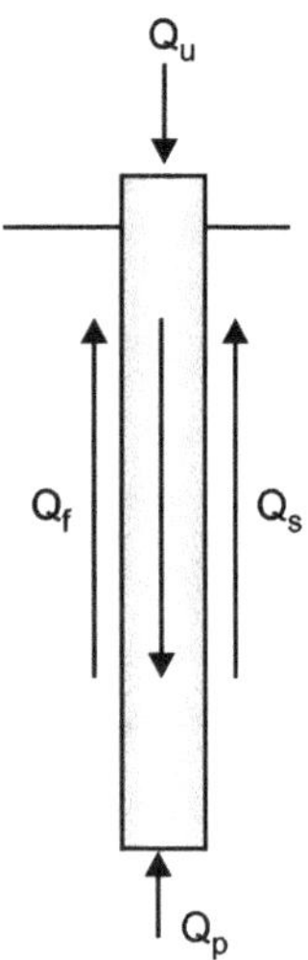

Fig. 5.3

- This is the ultimate skin friction resistance of the pile Q_f. It is only when the load at the top of the pile exceeds Q_f that the load in excess of Q_f begin to be transferred to the soil at the base of the pile.

- This load is known as point bearing load, which goes on increasing till soil at the base of the pile fails by punching shear failure. Load in bearing corresponding to this stage is called ultimate point load.

Value of N_c is taken as 9 and the value of N_q and D_c can be taken as below from the given below :

Table 5.2

State of Sandy Soil	Density Index	Value of N_q		Value of Critical Depth D_c
		Driven Pile	Bored Pile	
Loose	0.2- 0.4	60	25	6B
Medium	0.4 - 0.75	100	60	8B
Dense	> 0.75	180	100	15B

Value of K, δ and α can be taken as below

Table 5.3 : Values of K and δ

Pile Material	δ	Value of K	
		Loose Sand	Dense Sand
Steel	20^0	0.5	1.0
Concrete	0.75φ	1.0	2.0
Timber	$0.67\,\varphi$	1.5	4.0

Table 5.4 : Values of α

Consistency of Soil	N Value	α	
		Bored Pile	Driven Cast in Situ Pile
Soft to very soft	< 4	0.7	1.0
Medium	4 - 8	0.5	0.7
Stiff	8-15	0.4	0.4
Stiff to hard	> 15	0.3	0.3

- Value of α depends on undrained strength of the soil. Smaller the strength of the soil softer the consistency of soil and greater will be the tendency for the soil to adhere to the pile hence for such soil α will be close to 1.0 and for very stiff clays the adhesion coefficient will be as low as 0.3.

- Although value of α is less for stiff soil but still its contribution in skin friction is considerable due to its inherent high shear strength.

2. **Dynamic Method:**

- Dynamic pile driving formulas are based upon the theory that ultimate carrying capacity is equal to the ultimate driving resistance. These formulas are derived starting with the relation:

 Energy input = Energy used + Energy lost

- The energy used equals the driving resistance times the pile movement. Thus by knowing the energy input and estimating energy losses, driving resistance can be calculated from observed pile movements. Numerous dynamic formulas have been proposed. They range from the simpler Engineering News Formula, to the more complex Hiley Formula.

- **Engineering News Record Formula**

$$Q_U = \frac{W_h H}{S + C}$$

 C = 2.5cm for drop hammer,

  0.25cm for single and double acting hammer

- **Modified ENR Formula**

$$Q_U = \frac{1.25\,\eta_h E_h}{S + 0.25} \times \frac{W_h + e^2 W_p}{W_h + W_p}$$

- **Hiley's Formula**

$$Q_U = \frac{\eta_h \eta_b W_h H}{S + 0.5\,C}$$

where, η_b - Efficiency of hammer blow (ratio of energy after impact to striking energy of ram)

$$\eta_b = \frac{W_h + e^2 W_p}{W_h + W_p} \quad \ldots \text{for } W_h > eW_p$$

$$\eta_b = \frac{W_h + e^2 W_p}{W_h + W_p} - \left[\frac{W_h - e \cdot W_p}{W_h + W_p}\right]^2$$
$$\ldots \text{for } W_h < eW_p$$

where,

Q_u - Ultimate capacity of pile,

W_h - Weight of hammer in kg,

W_p - Weight of pile helmet and follower,

H - Height of drop of hammer in cm,

S - Penetration or set in cm per blow,

e - Coefficient of restitution,

C - Total compression = $C_1 + C_2 + C_3$,

C_1 - Temporary elastic compression of dolly and packing.

C_2 - Temporary elastic compression of pile,

C_3 - Temporary compression of soil,

η_h - Efficiency of hammer.

Efficiency of hammer can be taken as Table 5.5.

Table 5.5

Type	Efficiency %
Drop hammer	75 – 100
Single acting hammer	75 – 85
Double acting	85
Diesel hammer	85 – 100

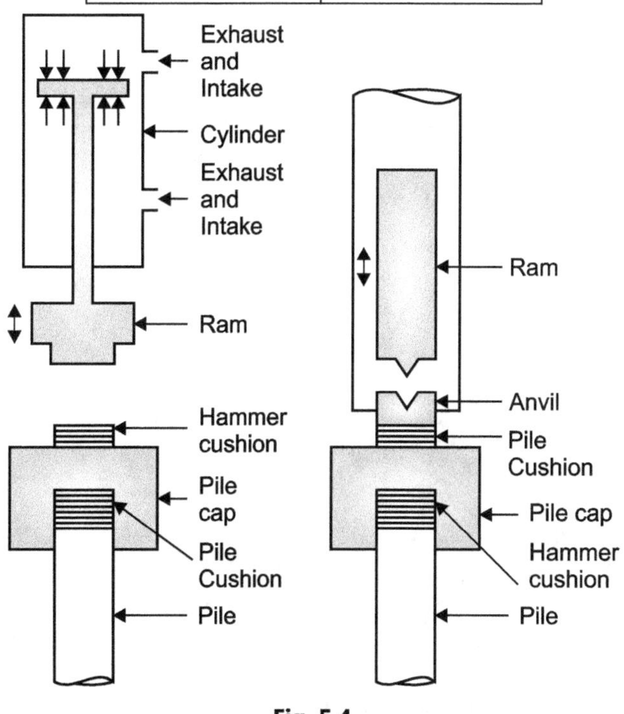

Fig. 5.4

3. **Pile Load Test**

- Actual load test on piles is the most effective and accurate method of determining the load capacity.

- The test is performed in one of the following ways.

  > **Continuous Loading :** Load applied is gradually increased.

  > **Cyclic Loading :** Load applied is removed and again applied (increased).

- The test is conducted on a test pile or working pile.

- The pile which is used only for testing (i.e. to find load carrying capacity); the same pile is not used for supporting the structure after the test, is called **'test pile'**.

- The pile which is used for testing and subsequently used for supporting structural loads is called **'working pile'**.

Fig. 5.5 : Pile load test setup

Continuous Loading Test

- Arrangement for the test is as shown in Fig. and is used to determine the settlement of pile at a given load and also to determine ultimate load which the pile can take at failure.

- Load is applied in equal increments of about 20% of estimated allowable load. Corresponding settlements are recorded with the help of three dial gauges symmetrically arranged over the test plate.

- Each load is kept for sufficient time till the rate of settlement becomes less than 0.02 mm/hr. Test piles are loaded until ultimate load is reached. Test load is increased upto a value of 2.5 times allowable load (estimated) or to a load which cause settlement equal to 10% of pile diameter whichever occurs earlier. The results are plotted in the form of load-settlement curve.

- Allowable load is determined from load settlement curve load.

  ➢ $\frac{1}{3}$ to $\frac{1}{2}$ of load which causes a settlement equal to 10% of pile diameter.

  ➢ $\frac{2}{3}$ of load which causes settlement of 12 mm.

  ➢ $\frac{2}{3}$ of load which causes a net settlement of 6 mm.

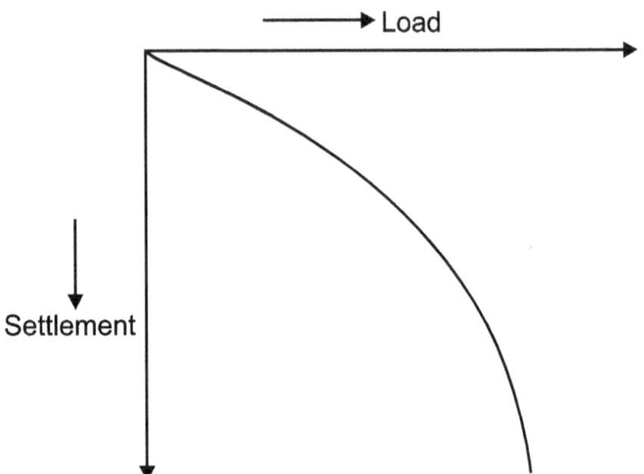

Fig. 5.6 : Load settlement curve

Cyclic Pile Load Test

- In this test, the load is raised upto a particular value then released to zero; again raised to higher value and released to zero. Settlements are recorded at each increment and decrement of load. This test is performed to separate out the point bearing load and frictional load.

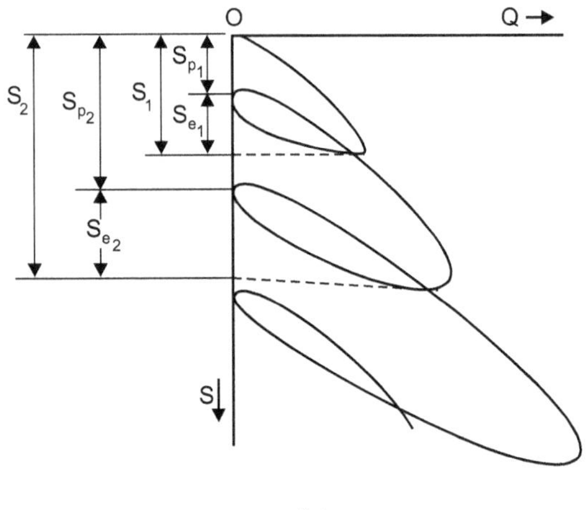

(a)

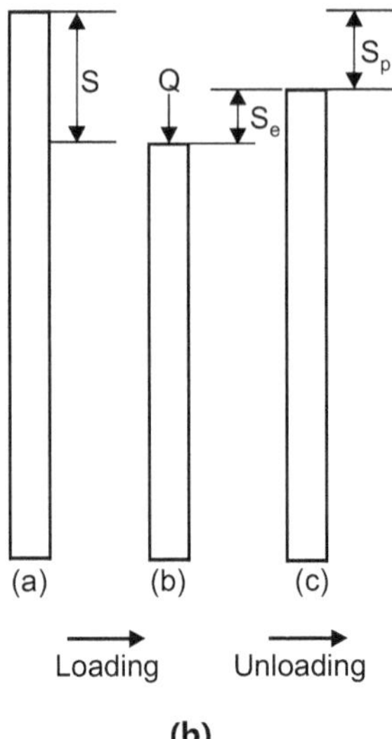

(b)

Fig. 5.7 : Elastic and plastic settlement of pile

- Fig. shows a pile which is being loaded. During loading the total settlement is "S" and then it is unloaded. While unloading, there is an elastic recovery 'Se' and residual settlement which is called 'plastic settlement'. If again we load the pile and unload it we will get elastic and plastic settlement which is as shown in graph. Thus, for each load we have elastic and plastic compression.

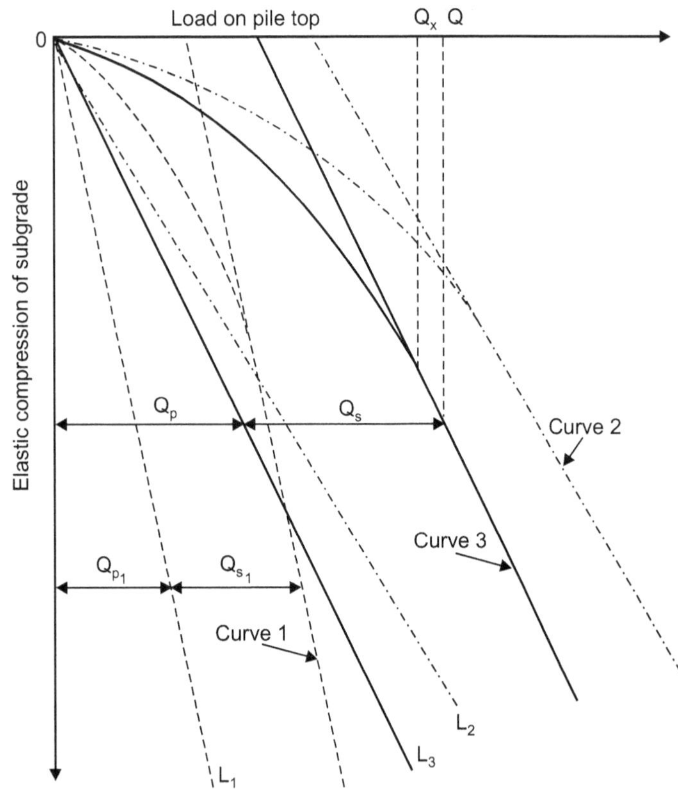

Fig. 5.8 : Separation of skin friction and point bearing

Procedure for Cyclic Pile Load Test :

1. Draw a graph between applied load and elastic settlement (Q_1, Se_1), (Q_2, Se_2) ... (Q_n, Se_n), as shown by curve '1' in graph.

2. Draw straight line "L_1" parallel to straight portion of curve 1 starting from origin.

3. This line approximately divides the load into Q_p and Q_s.

4. Find out skin load corresponding to each load. Say Q_{s1}, Q_{s2}, Q_{s3} ... Q_{sn}.

5. Find elastic compression of pile by formula for each load:

$$\Delta L = \frac{\left(Q - \dfrac{Qs}{2}\right) \cdot L}{A.E.}$$

  where, A : Cross-sectional area of pile

      E : Young's modulus of elasticity of pile material

6. Find modified values of elastic compression

$$\overline{S_e} = S_e - \Delta L$$

7. Plot graph between $\overline{S_e}$ and Q.

8. Repeat the procedure (steps 4, 5 and 6) till two subsequent values of Q_s are close to each other.

5.4 NEGATIVE SKIN FRICTION

- Negative skin friction is downward drag acting on the piles due to relative movement between the piles and the surrounding soil. When piles are driven through compressible soils, and site has newly placed fill or will be filled in the future, the possibility of negative skin friction should be investigated.

- Soft to medium clays, soft silt, peat, mud, etc are compressible soils. Lowering of ground water table in such compressible soils may also bring about negative skin friction. This occurs on the part of the shaft along which the downward movement of the surrounding soil exceeds the settlement of the pile.

- Negative skin friction could result from consolidation of a soft deposit caused by dewatering or the placement of fill. The dissipation of excess pore water pressure arising from pile driving in soft clay can also result in consolidation of the clay.

- Negative skin friction is calculated similar to positive skin friction,

$$Q_{ns} = Q_s = f_s A_s.$$

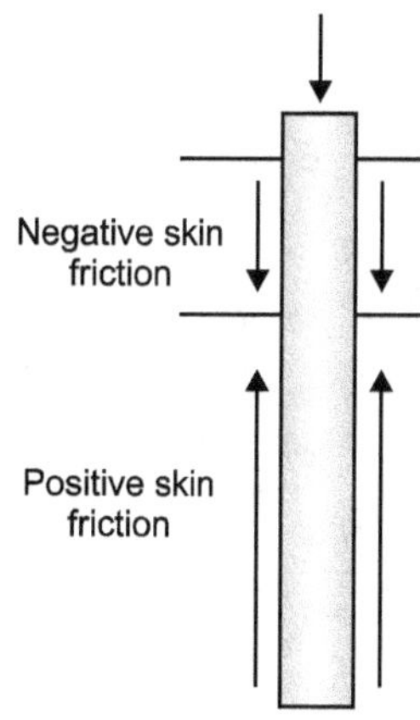

Fig. 5.9

- The magnitude of negative skin friction that can be transferred to a pile depends on (Bjerrum, 1973):

  - pile material,

  - method of pile construction,

  - nature of soil, and

  - amount and rate of relative movement between the soil and the pile.

- Development of negative skin friction on pile surface is not desirable, as it will impose additional load on the pile so pile when subjected to negative skin friction need to be designed for external load and negative skin friction, thus as far as possible such condition is avoided either by compacting the soil before driving the pile or the compressible soil is replaced by incompressible soil.

5.5 UNDERREAMED PILE

- An 'under-reamed' pile is one with an enlarged base or a bulb; the bulb is called 'under-ream'. There could be one or more under-reams in a pile; in the former case, it is called a single under-reamed pile and in the latter, it is said to be a multi-under-reamed pile.

- Under-reamed piles are cast-in-situ piles, which may be installed both in sandy and in clayey soils. The sides may be stabilized, if necessary, by the use of bentonite slurry, sometimes called 'drilling mud'. The under-reams are formed by special under-reaming equipment.

- The ratio of bulb size to the pile shaft size may be 2 to 3; usually a value of 2.5 is used. The bearing capacity of the pile increases because of the increased base area; the more the number of under-reams the more the capacity. Field tests indicate that an under-reamed pile is more economical than a straight bored pile for a given load.

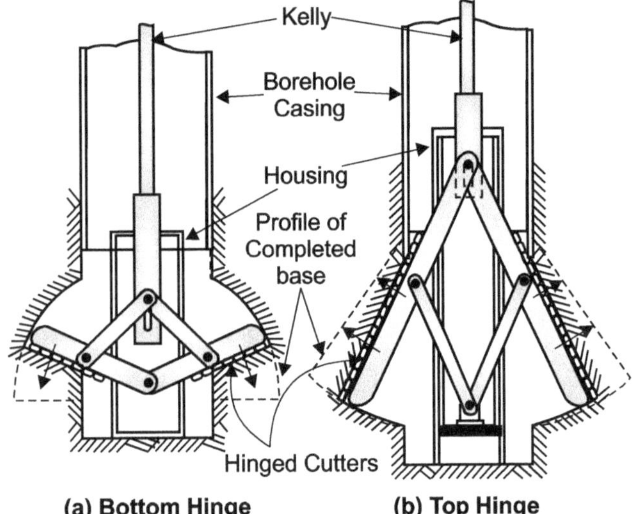

(a) Bottom Hinge　　　**(b) Top Hinge**

Fig. 5.10

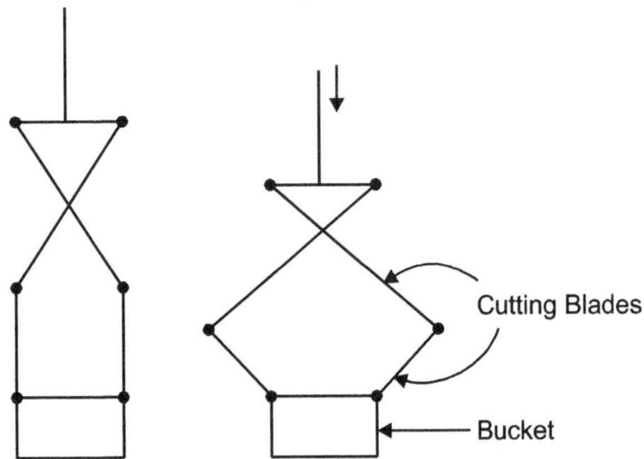

Fig. 5.11 : Underreaming tool in upright and expanded diameter position

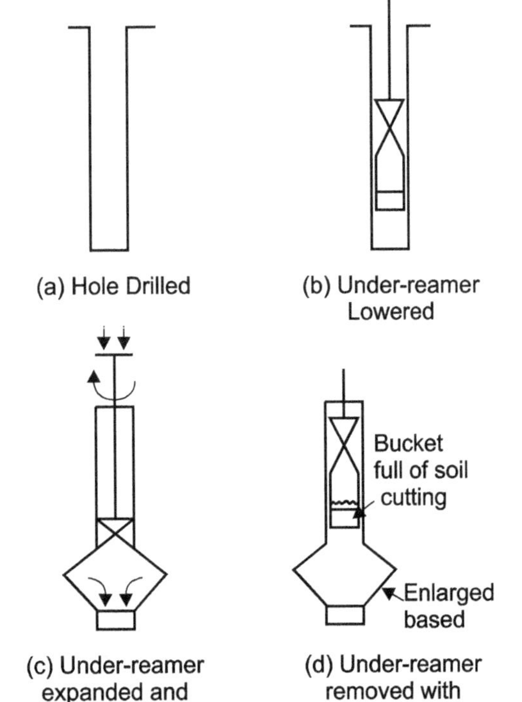

(a) Hole Drilled　　　(b) Under-reamer Lowered

(c) Under-reamer expanded and rotated　　　(d) Under-reamer removed with soil cutting

Fig. 5.12 : Construction of under-reamed pile

Procedure :

1. A bore is drilled by means of a hand operated auger up to required depth.

2. Under-reamer is lowered into the hole. It is then pressed down and rotated.

3. The blades widen under pressure and cut the soil from sides by rotation.

4. When the bucket is full, under reamer is pulled out and the soil is removed.

5. Tool is again lowered in to the hole and process is repeated.

Precautions:

- Verticality of the borehole need to be ensured by guiding frame.

- Casing pipe may be used in loose top soil.

- Obstructions to boring need to be broken down with cutting tool. In case obstruction cannot be handled, pile location need to be shifted.

- For boring in granular soil drilling mud is used to keep the side stable.

- Depth of borehole should be checked before inserting under-reamer otherwise the position of bulb will change.

- Spread should be checked before concreting.

- For underwater construction concrete of higher slump should be used by using tremie pipe.

- Proper compaction of concrete need to be done in bulb.

5.6　GROUP ACTION OF PILES

5.6.1 Pile Caps

- Structure is never supported on single pile. Generally, there will be a minimum of two or three piles under a foundation element or footing to allow for misalignments and other inadvertent eccentricities. Building codes may stipulate the minimum number of piles under a building element.

- Superstructure never rest directly on the pile, concrete slab is cast on the piles to form a base for supporting the structure such slab is called pile cap whose function is to receive the load from structure and transfer it to the piles.

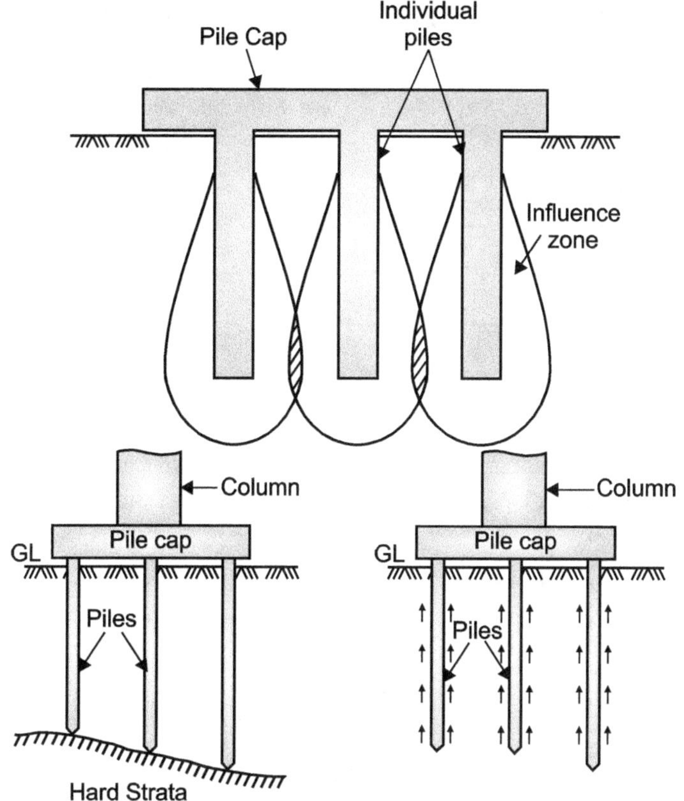

Fig. 5.13

5.6.2 Pile Groups

- Structure is never supported on single pile. Piles are always used in groups to support the structure. Generall, there will be minimum of two or three piles under a foundation element or footing to allow for misalignments and other inadvertent eccentricities.

- Building code may stipulate the minimum number of piles under a building element. Capacity of pile group is not necessarily equal to sum of capacity of individual piles, because installation and loading of pile bring about changes in
  - Stress condition
  - Consistency of clayey soil
  - Density index of granular soil

- Piles installed in a group to form a foundation will, when loaded, give rise to interaction between individual piles as well as between the structure and the piles.

- The pile-soil-pile interaction arises as a result of overlapping of stress (or strain) fields and could affect both the capacity and the settlement of the piles. The piled foundation as a whole also interacts with the structure by virtue of the difference in stiffness.

- This foundation-structure interaction affects the distribution of loads in the piles, together with forces and movements experienced by the structure.

- The analysis of the behaviour of a pile group is a complex soil-structure interaction problem. The behaviour of a pile group foundation will be influenced by:
  - Method of pile installation, e.g. displacement or replacement piles,
  - Dominant mode of load transfer, i.e. shaft friction or end bearing,
  - Nature of founding materials,
  - Three-dimensional geometry of the pile group configuration,
  - Presence or otherwise of a ground-bearing cap, and
  - Relative stiffness of the structure, the piles and the ground.

- The typical stresses/soil pressures produced from shaft friction or end bearing of a single pile and a group of piles are shown in Fig. 5.14. If piles are used in groups, there may be an overlap of stresses, as shown in Fig. 5.14 If the spacing is too close the overlap is large, and the soil may fail in shear or settlement will be very large.

- Though the overlapping zone of stresses obviously decreases with increased pile spacing, it may not be feasible since the pile cap size becomes too large and hence expensive.

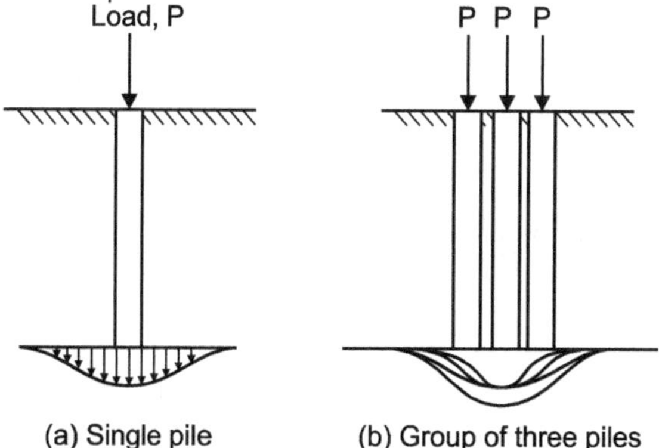

A: End Bearing Piles

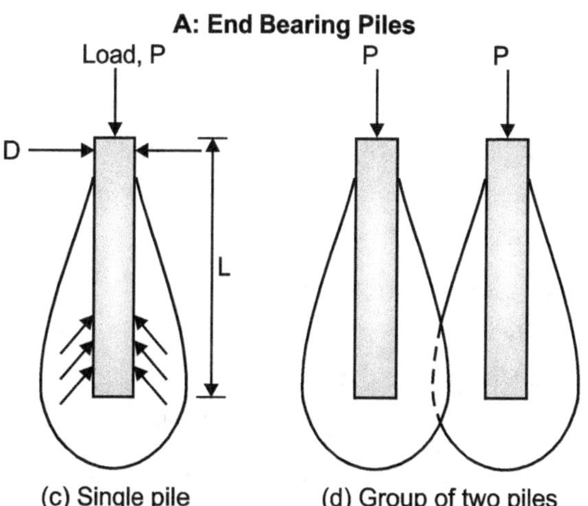

B: Friction piles

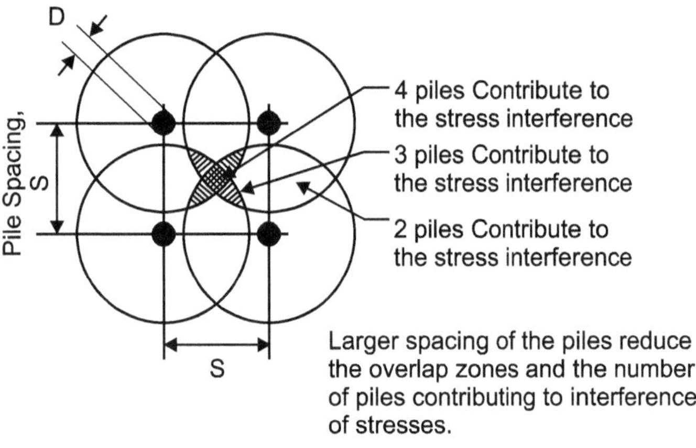

(e) Stress in soils due to piles groups

C: Interference of stresses in pile groups

Fig. 5.14

- Spacing of the piles should be such that neither it increases the cost of pile cap nor there is excessive overlapping of stresses. It depends on following factors

  ➢ Type of soil

  ➢ Method of installation

  ➢ Overlapping of stresses

  ➢ Efficiency of pile group

  ➢ Cost of foundation

- Optimal spacing will be 2.5D - 3.5D for vertical loads and may be suitably increased for lateral dynamic loads.

5.6.4 Efficiency of Pile Group

It is defined as,

$$E_g = \frac{\text{group capacity}}{\text{number of piles} \times \text{individual pile capacity}}$$

- If piles are driven into compressible cohesive soil or into dense cohesionless material underlain by compressible soil, then the ultimate axial compression capacity of a pile group may be less than that of the sum of the ultimate axial compression capacities of the individual piles. In this case, the pile group has a group efficiency of less than 1.

- In cohesionless soils, the ultimate axial compression capacity of a pile group is generally greater than the sum of the ultimate axial compression capacities of the individual piles comprising the group.

- In this case, the pile group has a group efficiency greater than 1. (granular soil densifies in the vicinity of a driven pile).

Table 5.6 : Effect of spacing of piles on group efficiency (AASHO Guidelines)

Pile Spacing c/c	Pile Group Efficiency	
	Clayey Soil	Sandy Soil
3D	0.67	0.67
4D	0.78	0.74
5D	0.89	0.80
6D	1.00	0.87
7D	1.00	0.93
8D or more	1.00	1.00

- **Empirical Equations to Calculate Efficiency of Pile Group :**

1. **Converse Lanarre's Formula**

$$E_g = 1 - \left\{\frac{m(n-1) + n(m-1)}{mn}\right\}\frac{\theta}{90}$$

2. **Seiler Keeney Formula**

$$E_g = \left\{1 - 0.479\left(\frac{S}{S^2 - 0.093}\right)\left(\frac{m+n-2}{m+n-1}\right)\right\} + \frac{0.3}{m+n}$$

3. **Los Angles Formula**

$$E_g = 1 - \{m(n-1) + n(m-1) + \sqrt{2}(m-1)(n-1)\}\frac{D}{\pi mnS}$$

where,   D : diameter of pile

S : Spacing of pile

m and n : number of rows or columns of piles

$$\theta : \text{Angle in degrees} = \tan^{-1}\left(\frac{D}{S}\right)$$

4. **Felds Rule :** Value of each pile is reduced by 1/16th on account of effect of nearest pile in each diagonal or straight row

5.6.5 Calculation of Capacity of Pile Groups : (Clayey Soil)

Capacity of the pile group can be calculated by considering group to fail by individual action and by block action

1. **By Individual Action :**

   Group capacity = no.of piles in group × capacity of single pile

2. **By Block Action**

   Group capacity = $P_g LC + A_g X9C - \gamma LA_g$

Above two values are compared and **whichever is least** is taken as actual capacity of pile group.

5.7 CAISSON FOUNDATION

The word caisson is derived from the French word **caisse** meaning a box. Caisson is a boxlike structure which is sunk through ground or water to remove water and soil during the process of excavation of foundation and subsequently becomes integral part of the substructure. It usually refers to a substructure element used in wet construction sites such as rivers, lakes, docks etc.

Classification of Caisson:

Caissons are classified into three types

1. Box caisson (Open at top and closed at bottom)

2. Pneumatic caisson (Open at bottom and closed at top)

3. Open caisson or Well (Open both at top and bottom)

Uses of Caissons:

- As foundation for bridge piers and abutments in rivers, lakes and seas, breakwaters and other shore works.

- In other protection works and large water front structures, such as pump house, which are subjected to huge vertical and horizontal forces.

- For large and multi-storey buildings and other structures, occasionally caissons are used.

5.7.1 Box Caisson

- This is a watertight vessel which is open at top and closed at bottom; the box could be made out of timber, RCC or steel. The box is constructed on the river bank then it is floated and towed to the position.

- It is then filled with D.L. and sunk at site where it serves as foundation floating caisson does not penetrate onto the soil. It simply rests on a hard, level surface; thus, the load-carrying capacity depends solely on the resistance at the base as there is no frictional resistance at the sides.

- A concrete cap is cast on its top to receive the loads from the superstructure. To prevent scour, rip rap is placed around the base.

Advantages :

- Since floating caissons are precast, good quality can be ensured.

- The installation of a floating caisson is quick and convenient.

- Floating caissons are less expensive than other types; they may also be transported at a low cost by floating.

- It can be used where construction of other types of caissons are not possible.

Disadvantages :

- The foundation bed has to be levelled before installing the caisson.

- The base of the caisson must be protected against scour.

- The load carrying capacity is smaller than that of other types of comparable size.

- This type is suitable only if a good supporting stratum is available at shallow elevation; otherwise, it becomes costly owing to deep excavation, as the saturated soil tends to flow into the excavation.

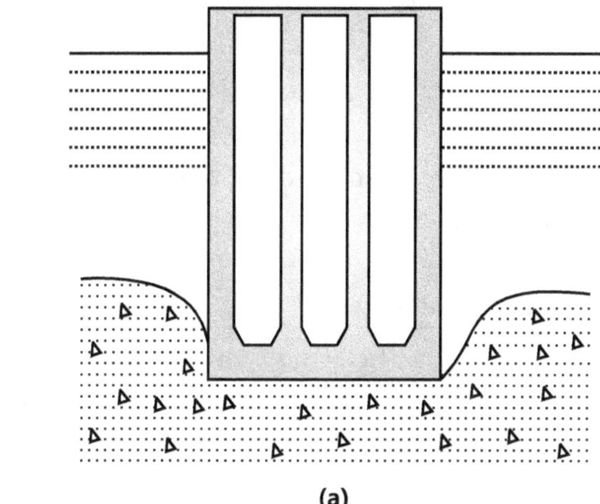

(a)

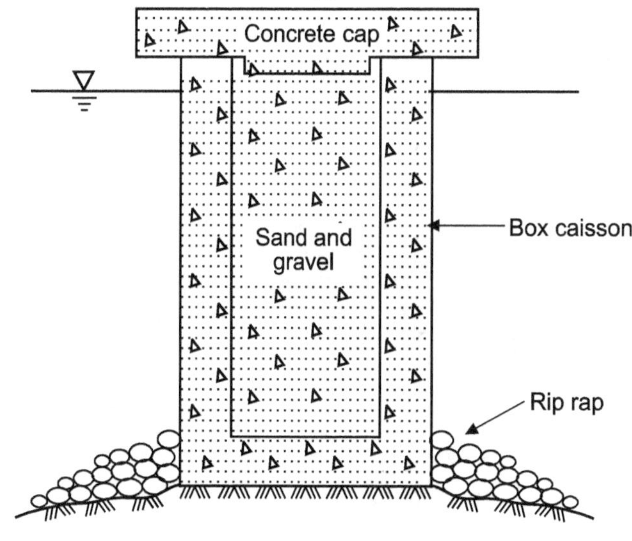

(b)

Fig. 5.15 : Box caisson

Construction:

Fig. 5.16

Procedure of Construction of Box Caissons :

1. Floating caissons are cast away from site on land.
2. Floated to desired location.
3. Caissons are filled with sand or gravel to facilitate sinking.
4. The desired base location is excavated and levelled as per requirement.
5. Caisson is then sunk to desired location and depth.
6. Rip-rap is provided.
7. Concrete cap is casted at the top to carry loads from superstructure.

Floating caissons are constructed by using RCC or steel. Internal strutting and diaphragm walls may be used if it is to be floated and placed in rough waters.

5.7.2 Pneumatic Caisson

Pneumatic caisson is open at bottom and closed at top between these two air pressure is maintained so that water does not enter the caisson so as to facilitate excavation and concreting.

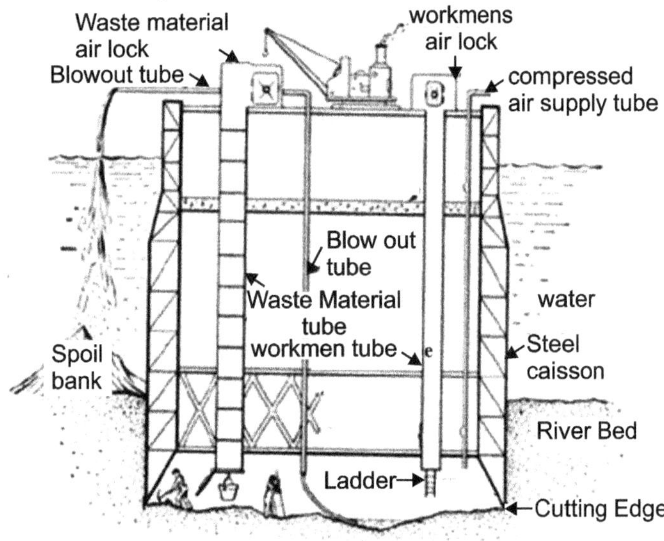

Fig. 5.17 : Pneumatic cassion

Component Parts of Caisson: A pneumatic caisson consists of the following component parts:

1. Working Chamber
2. Air Shaft
3. Air Lock
4. Miscellaneous Equipment

1.　Working Chamber:

- This is made of structural steel, about 3 m high, with a strong roof, and is absolutely air tight. The air pressure in the Chamber is raised above atmospheric and is kept at a certain specified value to prevent entry of water and soil into it.

- This pressure varies with the depth at which excavation is proceeding at any time. The outside surface is made smooth to reduce friction. A cutting edge is provided at the bottom to facilitate the sinking process.

- The air pressure must be sufficient to balance the full hydrostatic pressure due to water outside. However, there is a maximum limit to the air pressure in view of the physiological characteristics of human beings; a pressure greater than 0.4 N/mm^2 (400 kN/m^2) is beyond the endurance limit of human beings.

- Therefore, the maximum depth of water through which a pneumatic caisson can be sunk successfully is about 40 m. working under a pressure greater than 0.4 N/mm^2 may cause a special sickness called 'Caisson disease'.

2.　Air Shaft:

- This is a vertical passage which connects the working chamber with an airlock. It is meant to provide access to the working chamber for workmen. It is also used to transport the excavated material to the ground surface. In large caissons, separate shafts may be provided for men and materials.

- Air-shafts are made of Steel; the joints being provided with rubber gaskets to make them leak proof. Each Shaft is provided with its own air lock at its top. As the caisson sinks, the air shaft is extended to keep the airlock always above the water level.

3.　Air Lock:

- This is a steel chamber provided at the upper end of the air shaft above the water level. Its function is to permit the workmen to go in or come out of the Caisson without releasing the air pressure in the working chamber.

- The chamber of the airlock is provided with two air-tight doors, one of which opens to the shaft and the other to the atmosphere outside. When a workman enters the airlock through the outside door the pressure in the chamber is kept at atmospheric value. This is gradually raised till it becomes equal to that in the working chamber, and the workman allowed to enter the airshaft through the door to it, and to descend into the working chamber.

- The procedure is just reversed when one has to come out. However, the decompression must be effected much more slowly to prevent caisson disease. A minimum of half-an-hour is necessary for the pressure to be reduced from 0.3 N/mm^2 to atmospheric pressure. Fresh air is circulated into the working chamber by opening a valve in the airlock in order to prevent the air inside becoming stale.

- The workmen should not be made to work inside the working chamber under compressed air pressure for more than two hours at a stretch

4. Miscellaneous Equipment:

- Certain miscellaneous equipment such as motors, pressure pumps, and compressors are usually located outside at bed level.
- Pressure in the working chamber is maintained through an air pipe connected to a compressor. At least one stand-by unit for all equipment should be provided to cope with any emergency.

Advantages :

- Control over the work and foundation preparation are better, since all work is done in the dry.
- Obstruction from boulders or logs may be readily removed since direct visual inspection of the bottom near the cutting edge is possible.
- Concrete placed in the dry is more capable of attaining better quality and reliability.
- Plumbness of caisson is easier to control than with other types.
- Soil can be inspected, samples taken, and bearing capacity ascertained more reliably, if necessary, by *in-situ* testing.
- No settlement of adjoining structures need be apprehended since no lowering of ground water table is expected to occur.
- Large depths of foundation can be achieved to bed rock through difficult strata for major civil engineering works.

Disadvantages :

- Pneumatic caissons are highly expensive and hence should be used only when other types of caissons are not feasible.
- The depth of penetration is limited to 30 m to 40 m below water table.
- A lot of inconvenience is caused to the workmen while working under compressed air pressure, and they may be afflicted with caisson disease.
- Extreme care is required for the proper working of the system; even a small degree of slackness may lead to an accident.

Construction:

Pneumatic caissons may be:

1. Constructed at site or
2. Floated to the site and placed from barges.
3. The **sand island method** can also be used.

Procedure of Construction of Pneumatic Caisson:

1. Cutting edge is very carefully positioned to desired location.
2. Introduction of compressed air is needed into working chamber to keep off mud and water.
3. Working chamber is dewatered and made dry.
4. Workmen are allowed to enter into shafts through air locks.
5. As workmen carryout excavation, caisson starts sinking.
6. The air pressure is raised to make up with the pressure due to the head of water as the sinking goes on.
7. Excavated material is transferred to the top with the help of buckets through shafts.
8. Blow-out pipes can be used in case of granular soils to transfer excavated material.
9. When caisson reaches its final and desired position, working chamber is filled with concrete.
10. Till concrete gets hardened, the air pressure in the chamber is kept constant.
11. Finally, shafts are also filled with concrete after dismantling shaft tubes.

Caisson Disease:

- In case of sinking process of pneumatic caisson, workers or workmen have to work in working chamber under compressed air. If the compressed air pressure is more than 0.35 N/mm^2 to 0.4 N/mm^2, then workmen may suffer from the following pains:

  ➢ Workmen may suffer from giddiness.

  ➢ There is a pain in ears of workmen.

  ➢ There is breaking of ear drums of workmen.

  ➢ There is bursting of blood vessels in the nose or ears of workmen.

- The above mentioned pains are not that serious or fatal, but workmen are actually suffering during decompression and effect causing depression is called caisson disease the symptom of the disease were dizziness, double vision, headache, trouble to speaking, pains in legs etc. The disease results in the loss of consciousness, paralysis or sometimes even death.

- Excessive oxygen gets absorbed in the blood and tissues during decompression is more troublesome to workmen. Absorbed oxygen gas is thrown out of blood in the form of bubbles which can block in vessels and may cause bursting of vessels.

- If bubbles are developed in joints it causes bends

- If the bubbles are developed in spinal cord, it causes paralysis and if the bubbles are developed in heart, it causes heart attack.

Precautions to Avoid Caisson Disease:

- Person should not work continuously for more than four hours per day.
- Temperature in the working chamber should be roughly around 25° C.
- Person with strong heart, relatively normal blood pressure and good circulation should be employed.
- Worker should undergo complete check up from a physician once in a month.
- Rate of compression and decompression should be gradual without causing any discomfort.

Fig. 5.18 : Working chamber of pneumatic caissons

5.8 OPEN CAISSON OR WELL

Open caisson is open both at the top and the bottom during construction. It is provided with a cutting edge at the bottom to facilitate sinking. When the caisson has reached the desired location, a fairly thick concrete seal is provided. The thickness of the seal may range from 1.5 to 4.5 m.

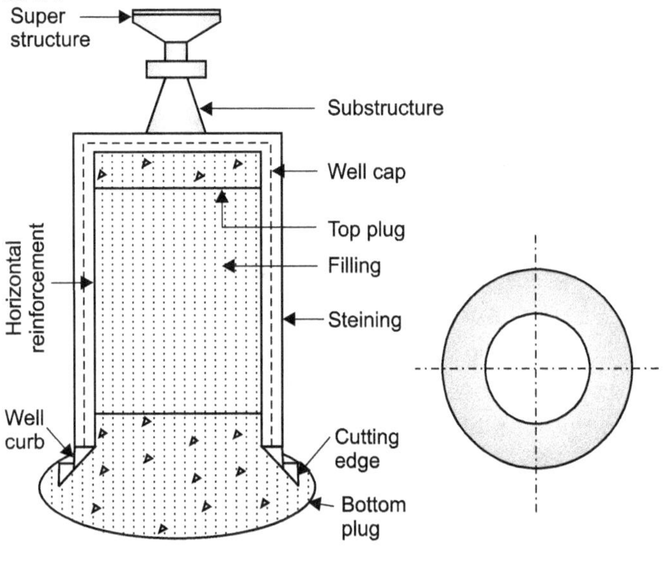

(a) Vertical section   **(b) Horizontal section**

Fig. 5.19 : Open caisson

Advantages :

- It is feasible to extend to large depths.
- The cost of construction is relatively low.

Disadvantages:

- The bottom of the caissons cannot be inspected and thoroughly cleaned.
- The concrete seal is necessarily placed in water, and may not be satisfactory.
- The help of divers may be necessary for excavation near the haunches at the cutting edge.
- If obstructions of boulders or logs are encountered, the work is slowed down significantly.

5.8.1 Component Parts of a Well Foundation

1. Cutting edge
2. Curb
3. Concrete seal or Bottom Plug
4. Steining
5. Top Plug, and
6. Well Cap

1. **Cutting Edge :** The function of the cutting edge is to facilitate easy penetration or sinking into the soil to the desired depth. As it has to cut through the soil, it should be as sharp as possible, and strong enough to resist the high stresses to which it is subjected during the sinking process. Hence it usually consists of an angle iron with or without an additional plate of structural steel

2. **Curb:** The well curb is a transition member between the sharp cutting edge and the thick seining. It is thus tapering in shape. It is usually made of reinforced concrete as it is subjected to severe stresses during the sinking process.

3. **Concrete Seal or Bottom Plug:** After the well foundation is sunk to the desired depth so as to rest on a firm stratum, a thick layer of concrete is provided at the bottom inside the well, generally under water. This layer is called the concrete seal or bottom plug, which serves as the base for the well foundation. This is primarily meant to distribute the loads on to a large area of the foundation, and hence may be omitted when the well is made to rest on hard rock.

4. **Steining :** The steining forms the bulk of the well foundation and may be constructed with brick or stone masonry, or with plain or reinforced concrete occasionally. The thickness of the steining is made uniform throughout its depth. It is considered

desirable to provide vertical reinforcements to take care of the tensile stresses which might occur when the well is suspended from top during any stage of sinking.

5. **Top Plug:** After the well foundation is sunk to the desired depth, the inside of the well is filled with sand either partly or fully, and a top layer of concrete is placed. This is known as 'top plug'.

6. **Well Cap:** The well cap serves as a bearing pad to the superstructure, which may be a pier or an abutment. It distributes the superstructure load onto the well steining uniformly.

Procedure of Construction of Open Caisson:

1. Construction of well curb.
2. Construction of well steining.
3. Sinking process.
4. Sand filling.

1. Construction of Well Curb:

- In case of a dry river bed the well curb is to be built is placed at the correct position after excavating the bed for about 150 mm for seating. Sand island is constructed if water is present up to the depth of 5 m.

- Wooden sleepers can be placed below cutting edge to distribute load evenly. The shuttering of well curb is erected, then reinforcement for the curb is placed in position. Concreting of curb is done without gap. Curb is allowed to cure for 7 days and after 7 days shuttering and sleepers are removed.

2. Construction of Well Steining :

- The well steining is constructed with the height of 1.5 m at a time. Sinking process is continued after concrete is set for 24 hrs. When well reaches 6 m depth below ground, then steining can be raised 3 m at a time.

3. Sinking Process:

- It is an important step in the construction of well foundation, sinking process is started when curb is cast and first stage of steining is ready after curing.

- The area at which the well is to be sunk shall be excavated to the approximate top elevation of the completed well foundation, In case spring level is higher than the top elevation of the completed well, the excavation shall be maintained in a dry condition by utilizing cofferdams and un-watering methods.

- The well curb shall be constructed in place in the proper position, or if a pre-cast curb is used it shall be set in proper level position at the surface below which well sinking is to be carried. If a precast curb is used it shall be set in proper level position. Precast curb shall not be handled until they have been cured for at least 14 days. The vertical bars for masonry reinforcement (wherever reinforced brick work has been specified or specially directed by the Engineer-in-Charge) shall be attached to the steel angle cutting edge by means of nuts and washers

- After placing of precast curbs; or not earlier than three days after placing of Concrete for cast-in-place curbs, brick Masonry walls with an approximate height of 1.5 meters shall be constructed on the curbs

- The well shall be sunk by dredging (defined as removal of material from inside the well) until the top of the masonry is approximately 0.7 meters above the ground, whereupon straightedges for another lift of masonry shall be set and a second lift of masonry up to 3 meters in height constructed. Care shall be taken that the exterior faces of succeeding masonry lifts are constructed parallel to the axis of the well rather than plumb so as to indicate any uneven sinking of the well and permit remedial action to be taken. Each successive lift of masonry shall be allowed to set for three days before sinking of the well is resumed.

- In case spring level is high, then operation shall be maintained in a dry condition using cofferdams and dewatering methods.

- Dredging shall be accomplished in such a manner that the hole within the well shall not be extended below the cutting edge by more than 1.25 meter and that, when the well is sunk to its final position, the material outside of the well will not have been disturbed.

- Each well shall be frequently checked for plumb by means of plumb lines and mason's level or other approved means. Corrective action, consisting of dredging from the high side until the well rights itself, shall be taken immediately if the well is found to be sinking unevenly. If required, weights shall be added at the top of the well masonry on the high side. The corrective force shall be applied concurrently with sinking of the well.

- Each well shall also be frequently checked for longitudinal and lateral drift during sinking by the use of a suitable sighting device.

- If the well does not sink as the dredging is advanced, a greater height of masonry, weighting or running shall be employed. Running, defined as the practice of removing water from within the well to reduce buoyancy and thereby increase the effective weight of the well.

- When a well has been sunk to its indicated elevation, the bottom interior of the well shall be carefully sounded to detect the presence of any material within the space to be occupied by the bottom plug, and any material so detected shall be removed. The false masonry if constructed to increase the weight for sinking shall also be removed.

- Without un-watering, the bottom plug concrete shall be placed by means of a tremie.

4. Sand Filling:

- After the bottom plug concrete has set at least one day, and without un watering the well, the sand filling shall be placed in lifts not exceeding one meter in depth and with a 24-hour elapse of time between placement of the lifts to permit settlement of the fill.

- After the sand filling is placed to the demarcated elevation within the well, the top plug concrete shall be placed and screeding level at the elevation of the top of the brick masonry.

- If the well is titled within the permissible limits, the masonry at the top of the well shall be constructed, so that the top surface of the masonry around the perimeter of the well meets the specified elevation.

5.8.2 Shapes of Well

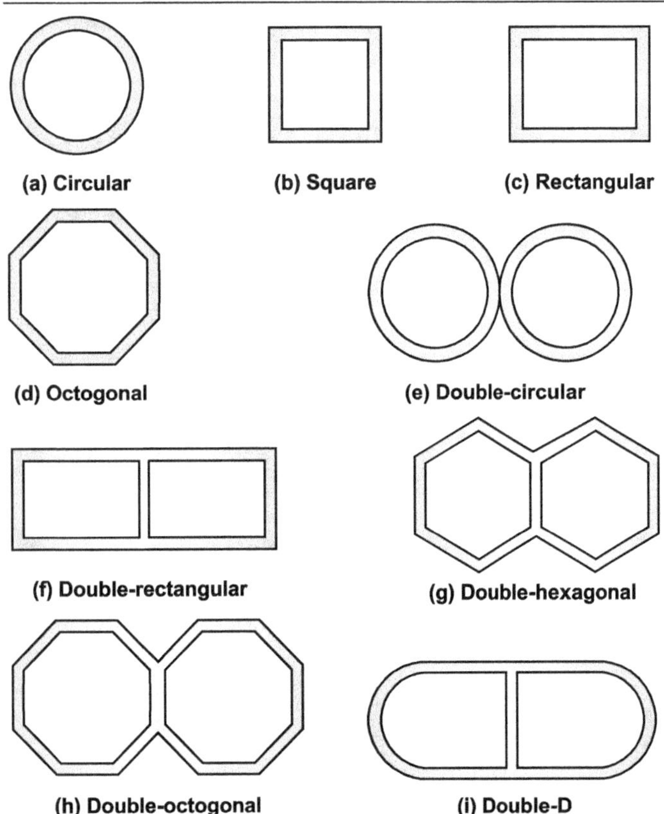

Fig. 5.20

Circular Well:

- It is the simplest to construct, easy to sink and has uniform strength in all directions. Simple to correct the tilt for circle. Only disadvantage for this shape is the limitation of its size which restrict its use to bridge with smaller pier.

Twin Circular Well:

- Two independent wells placed close to each other and provided with a common well cap over which the pier can be built. Spacing between wells and its diameter can be decided based on the width and length of pier. Both the wells have to be sunk together to avoid possibility of tilt and shift.

Double D – Well:

- This is the most common type of foundation in major bridges with multiple lane. This shape facilitates easy casting and sinking

5.8.3 Grip Length

- A well foundation should be sunk below the maximum scour depth such that there is adequate lateral stability. The depth of the bottom of the well below the maximum scour level is known as the 'Grip Length'.

- The scour depth can be ascertained by one of the following approaches:

  ➤ Actual sounding at or near the proposed site immediately after a flood, at any rate before there is any time for silting up appreciably.

  ➤ Theoretical methods taking into account the characteristics of flow like the direction, depth, and velocity, and those of the river bed material.

- In case the first approach of taking soundings is not feasible, the second approach may be used and the normal depth of scour may be calculated by Lacey's formula

$$d = 0.473 \left\{ \frac{Q}{f} \right\}^{1/3}$$

where, d : normal scour depth, measured below high flood level (m),

  Q : design discharge (m^3/s),

  f : Lacey's silt factor

Lacey's silt factor is given by

$$f = 1.76 \sqrt{d_m}$$

where, d_m : mean diameter of the particles (mm).

- The grip length for wells of railway bridges is taken as 50% of maximum scour depth, generally, while for road bridges 30% of maximum scour depth is considered adequate.

- According to IS: 3955-1967, the depth should not be less than 1.33 times the maximum scour depth. The depth of the base of the well below the scour level is kept not less than 2 m for piers and abutments with arches, and 1.2 m for piers and abutments supporting other types of structures.

Note: Maximum scouring depth can be taken as twice of normal scour depth.

Table 5.7 : Comparison of Pneumatic Caisson and Open Caisson

Pneumatic Caisson	Open Caisson
Labour cost is more.	Labour cost is low.
Excavation can be possible in any type of soil.	Excavation in hard strata is difficult.
Foundation can extend to max depth of 40m below water surface.	No limit for depth of foundation.
Special compressed air is needed in working chamber for dredging.	No compressed air is required.
High risk to workers.	No risk for worker.
Bottom of this caisson is cleaned in dry condition.	Cleaning is done under water.
Special precautions are required to avoid disease.	No precautions are required.
Obstacles can easily be observed and tackled during excavation.	It is not possible to see the obstacles.
Direct access to bottom of caisson for manual excavation.	No direct access to bottom of caisson.

5.8.4 Tilts and Shifts of Well

- Well normally should be sunk straight and at the correct position. However, it is not an easy task some displacement is bound to be there (translational and or rotational).

- Shift of well represent overall lateral displacement of wells axis from its true position in plan; and tilt represent rotation of a well about its bottom. These displacement results in changing of span length of the bridge and eccentric loading on pier.

Causes of Tilt:

- Non-uniform soil condition.

- Presence of large boulder or obstruction.

- Quick condition.

- Sudden sinking (blasting).

- Method of sinking.

Precautions to Avoid Tilt:

- Outer surface of well curb and steining should be regular and smooth.

- Diameter of curb should be kept about 4 – 8 cm larger than outer diameter of steining and well should be placed symmetrically.

- Cutting edge of curb should be of uniform thickness and sharpness.

- Dredging should be done uniformly on all sides in a circular well and in both pockets in case of twin wells.

Rectification of Tilt :

I.S. 3955-1967 recommends that tilt should be generally limited to 1 in 60 and shift should be restricted to 1% of depth of sunk. In case if these limits are exceeded then following measures need to be taken to rectify the tilt :

1. Regulation by Excavation
2. Providing Temporary Obstacles below the Cutting Edge
3. Eccentric Loading
4. Water Jetting
5. Pushing the Caissons or Well with Jack
6. Pulling the Well or Caisson
7. Strutting the Caissons or Well

1. **Regulation by Excavation:** Sinking of caisson on higher side due to excess excavation is more. This is all right in the early stages, otherwise dewatering of caisson or well is needed and open excavation may be done on higher side.

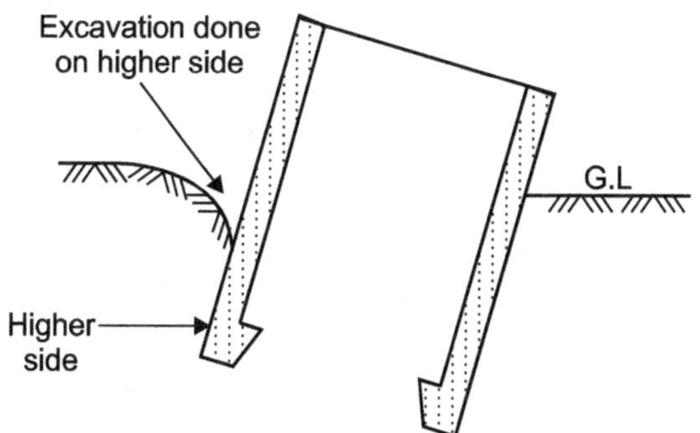

Fig. 5.21 : Regulation by excavation

2. **Providing Temporary Obstacles below the Cutting Edge:** Rectification of tilt can be done by inserting the wooden sleeper temporarily as an obstacle below the cutting edge on the lower side so as to prevent further tilt of the well or caisson.

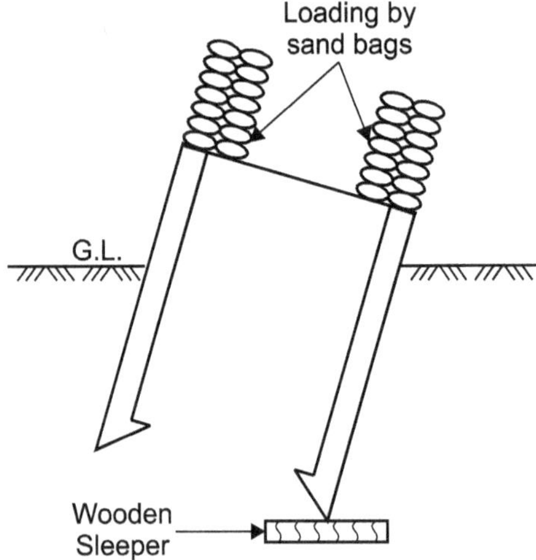

Fig. 5.22 : Regulation by providing temporary obstacles

3. **Eccentric Loading :** The caisson is normally given the additional loading called kentledge in order to have necessary sinking effort. In this method, eccentric loading or kentledge is applied in higher side so as to have greater sinking effort For proper application of eccentric loading a platform with projection on higher side can be placed over the top of caisson •The eccentric load is kept on projected part of platform. Thus tilt can be rectified

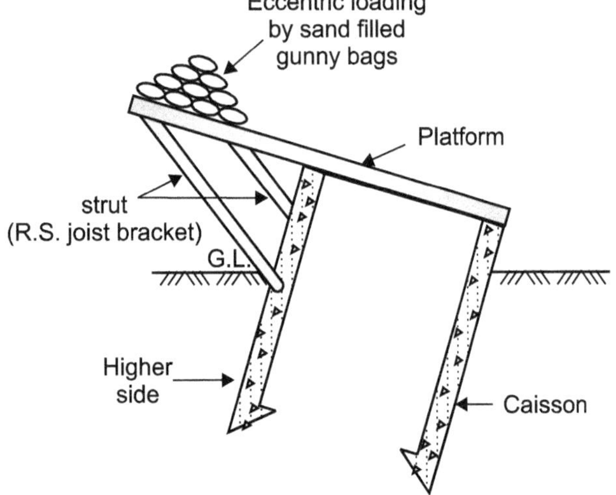

Regulation by eccentric loading

Fig. 5.23

4. **Water Jetting :** This is the one of the method used to prevent tilting. In this method, water jet is forcedly applied on tilt.

Application of water jetting on higher side reduces skin friction. Thus the tilting is rectified. This method is not more effective but gives the better result if used with the combination of other methods

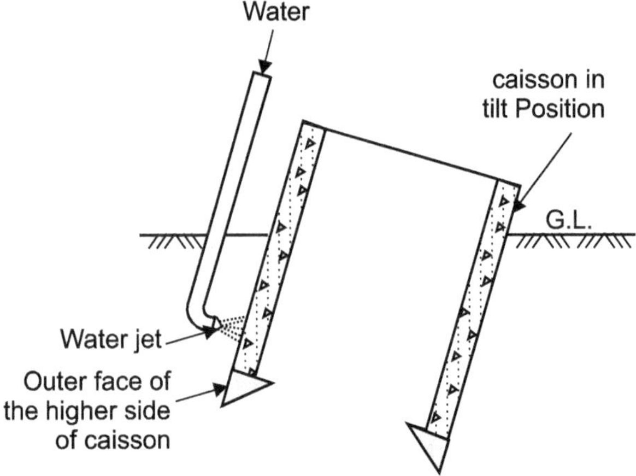

Fig. 5.24 : Regulation by water jetting

5. **Pushing the Caissons or Well with Jack:** Mechanical jack or hydraulic jack can be used to rectify the tilt of well or caisson. Well or caisson can be pushed by jack to bring it to a vertical position.

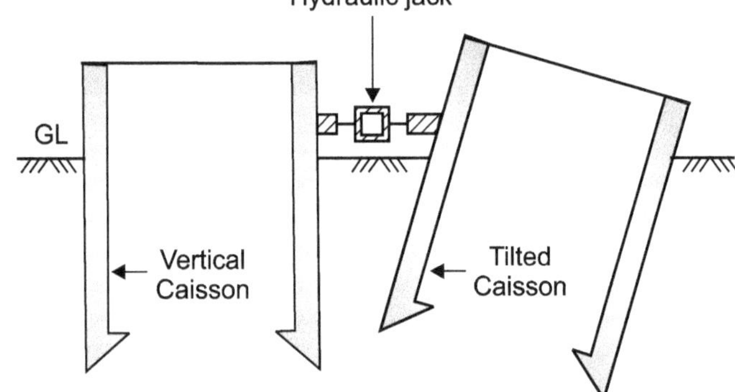

Fig. 5.25 : Regulation by pushing action

• **Pulling the Well or Caisson:** This method is most suitable and effective in preliminary or early stages of sinking operation. Steel ropes or cables are used pull the caisson or well. Pulling of caisson or well is done on higher side of well or caisson

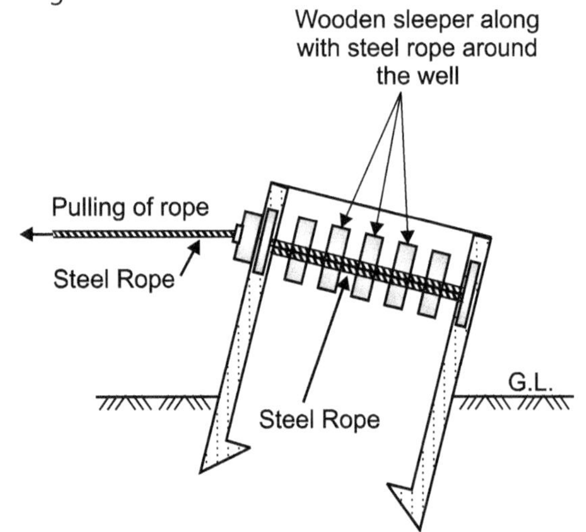

Fig. 5.26 : Regulation by pulling action

7. **Strutting the Caissons or Well :** Method of strutting the caisson or well is used to prevent any further and possible rise in tilting of the caisson or well. The caisson or well is supported on the tilting side by giving inclined support of a strong wooden member. This inclined wooden member is called as a strut.

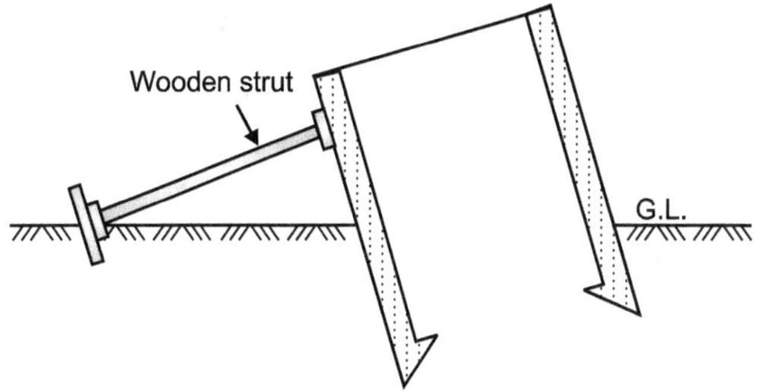

Fig. 5.27 : Regulation by strutting

Advantage and Disadvantage of Caisson Foundation Compared to Pile Foundation

Advantages :

- It can be carried through layers containing large boulders through which pile cannot penetrate.
- Does not produce vibration, heaving thus less damage to adjacent structures.
- Cost of machinery as well as noise level are low compared to that in case of driven piles.
- Foundation layer can be inspected and tested physically and hence reliability of design is good.
- It can sustain larger lateral force and offer more effective resistance to destructive forces due to floating objects and scour.

Disadvantages :

- For structure on land of medium size caisson may be more expensive than pile.
- Excavation of caisson in granular soil below water table is difficult.
- Caving in of the soft soil may cause problems.
- Seating of caisson on an irregular surfaces of bed rock is difficult.

Forces Acting on the Well Foundation: Well foundation will be subjected to following forces

- Dead load
- Live load
- Impact load
- Wind load
- Water pressure
- Longitudinal force (tractive force/ braking force)
- Earth pressure
- Centrifugal force
- Buoyant force
- Seismic force
- Temperature stress

5.9 SHEET PILE

- Sheet pile wall is a common type of earth retaining structure made of individual sheet piles driven in the ground. It consists of number of sheet piles driven side by side into the ground, thus forming a continuous vertical wall. These walls are commonly used for:

  1. Waterfront construction.
  2. Temporary construction such as cofferdam.

- These walls are **not** suitable in following situations:

  1. When height of wall is more due to high flexural stresses.
  2. If sub-soil is rocky strata, it prevents penetration of pile (pile derives its stability from depth of embedment.)

- Sheet piles are prefabricated members made of timbers, R.C.C. or steel.

- Timber sheet pilings are used for light lateral loads (low height), commonly used for braced excavation.

- R.C.C. sheet pilings are precast concrete members, these members should be designed for handling and driving stress due to their heavy weights and also displace large volume of soil, hence are used rarely.

- Steel sheet piles are most common type of piles that are being used and consists of structural members with interlocking either 'Finger and thumb' type or 'Ball and socket type' to engage with one another. These sheet piles have following advantages over other.

  ➢ It is relatively light weight.

  ➢ It may be reused several times.

  ➢ It has high resistance to driving stress (in hard strata).

  ➢ More service life.

  ➢ Easy to alter the length of pile (welding/bolting).

Classification of Sheet Pile: Based on the way it resists the load sheet pile walls are classified into two types

1. Cantilever sheet pile and
2. Anchored sheet pile walls.

In case of cantilever sheet pile wall lateral load coming on it is resisted by cantilever action. In case of anchored sheet pile wall the load will be resisted by cantilever action along with anchorage to minimize lateral movement.

1. Cantilever Sheet Pile

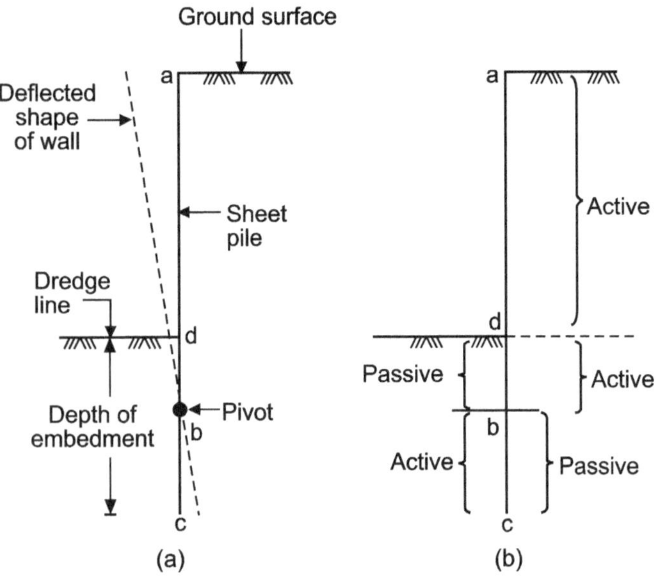

Fig. 5.28

- Cantilever sheet pile wall derives its stability entirely from lateral resistance of soil into which it is driven. The sheet pile wall is adequately embedded into the soil below the dredge line.

- These walls resist the load by cantilever action hence are suitable only for moderate heights because bending moment increases with the cube of cantilever height of wall.

- Fig. 5.28 shows cantilever sheet pile and its deflected shape under the action of earth pressure. Portion 'ab' of wall is subjected to only active earth pressure. Portion 'db' is subjected to active pressure from one side and passive pressure from other side and portion 'dc' is subjected to passive pressure from one side and active from other side.

- Determination of depth of embedment of sheet pile for resisting the lateral pressure is called design of sheet pile wall for which following cases are considered :

  ➢ **Case 1 :** Cantilever sheet pile wall in sandy soil above and below dredgeline.

  ➢ **Case 2 :** Cantilever sheet pile wall in clayey soil above and below dredgeline.

  ➢ **Case 3 :** Cantilever sheet pile wall with sandy soil above dredgeline and cohesive soil below dredgeline.

Case 1: Sandy Soil Above and Below Dredge Line :

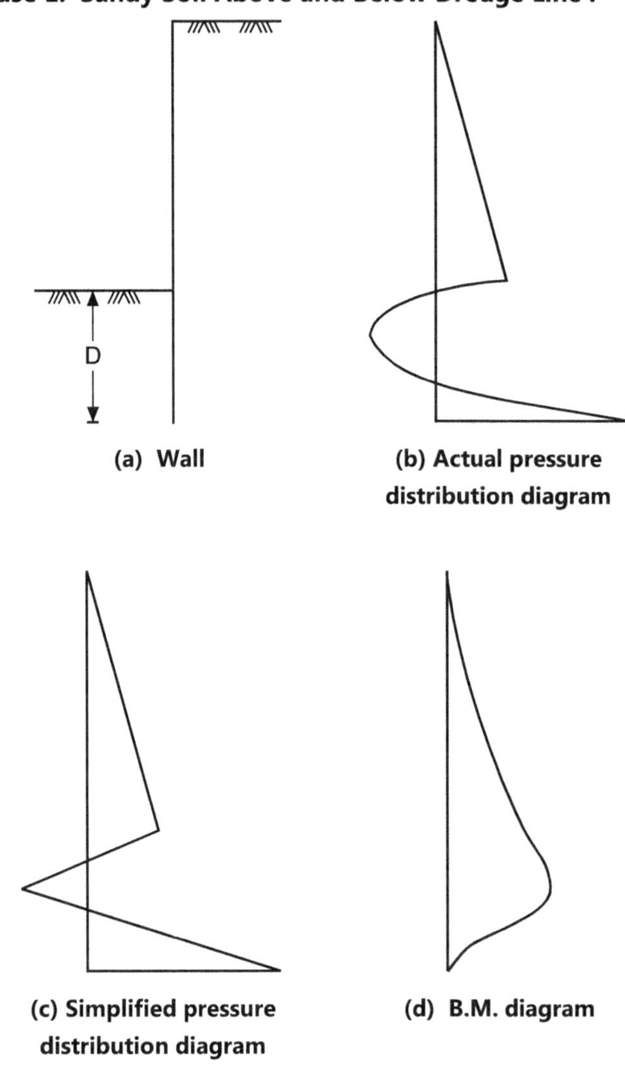

Fig. 5.29

Fig. 5.29 shows the sheet pile wall, active pressure distribution diagram acting on the wall, simplified pressure distribution diagram (which is used for design purpose) and bending moment diagram.

Depth of Embedment by Approximate Analysis : In approximate analysis the active earth pressure distribution on the back and passive earth pressure distribution on the front and concentrated passive pressure at the base of wall is considered for simplicity in analysis.

Taking moment about base, (Refer Fig. 5.29)

$$P_A \left(\frac{H + D}{3} \right) = P_p \cdot \frac{D}{3}$$

$$\frac{1}{2} K_a \cdot r (H + D)^2 \cdot \left(\frac{H + D}{3} \right) = \frac{1}{2} K_p \cdot r \cdot D^2 \cdot \frac{D}{3}$$

Above equation is solved for 'D' and then obtained value of 'D' is increased by 20-40%.

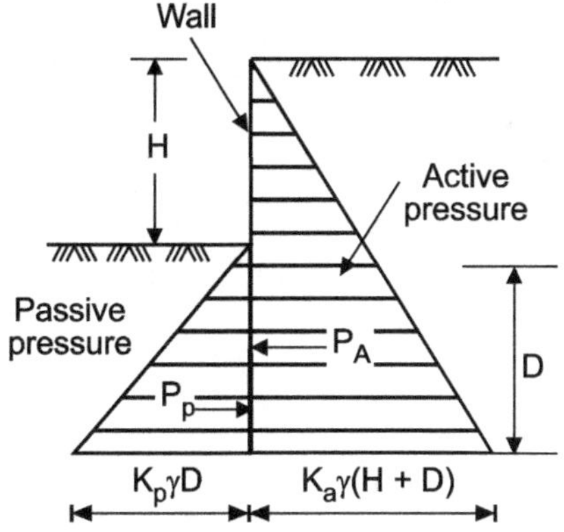

Fig. 5.30 : Pressure distribution diagram for approximate analysis

Case 2: Cantilever Sheet Piles in Cohesive Soil:

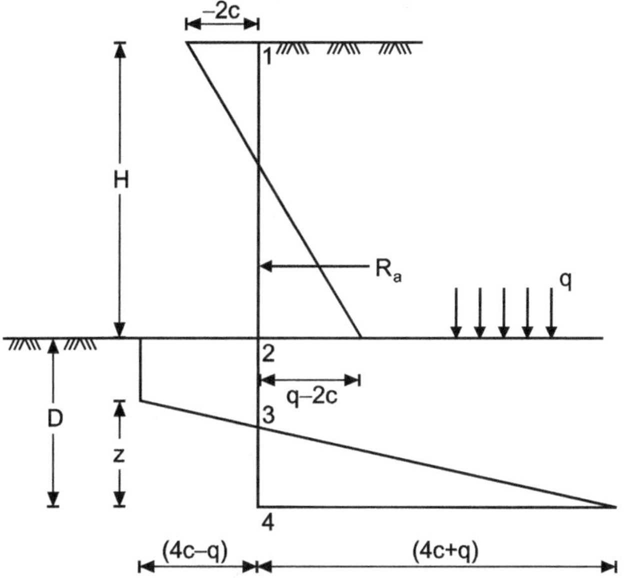

Fig. 5.31 : Pressure distribution diagram

Fig. 5.31 shows the pressure distribution diagram for sheet pile wall embedded in cohesive soil and backfilled with cohesive soil.

Active and passive pressure in general are given by,

$$P_a = \sigma_v K_a - 2c\sqrt{K_a}$$

and
$$P_p = \sigma_v K_p + 2c\sqrt{K_p}$$

For purely cohesive soil $\varphi = 0$, thus earth pressure coefficient = 1 (active and passive)

Thus $P_a = \sigma_v - 2c$

and $P_p = \sigma_v + 2c$

At point 1, vertical stress is $\sigma_v = 0$ thus

$$p_1 = 0 - 2c = -2c$$

At point 2, to the right (just above 2),

$$\sigma_v = rH = q;$$

thus, $p_{2r} = q - 2c$

At point 2, to the left (just below 2),

$$\sigma_v = 0;$$

thus $p_{21} = 0 + 2c$

Net pressure at point 2

$$= \text{Passive pressure } (p_{21}) - \text{Active pressure}(p_{2r})$$
$$= 2c - (q - 2c) = 4c - q$$

At point 4,

Net pressure = Passive pressure – Active pressure

$$= (q + rD) + 2c - (rD - 2c)$$
$$= 4c + q$$

To determine the depth of embedment, we apply equilibrium equation.

We neglect the –ve pressure for calculation.

$$\sum H = 0$$

$$R_a + [(4c + q) + (4c - q)]\frac{\bar{z}}{2} - (4c - q)D = 0$$

$$\sum M = 0 \text{ about base}$$

$$R_a (D + \bar{y}) - \frac{D^2}{2}(4c - q) + [4c - q + 4c + q]\frac{\bar{z}}{3} \cdot \frac{\bar{z}}{2} = 0$$

Solving above equations, we get the embedment depth. Then increase the depth by 20-40%.

Case 3 : Granular Soil Above and Cohesive Soil Below:

At point 1, $\sigma_v = 0,$

$$p_1 = \sigma_v K_a - 2c\sqrt{K_a}$$
$$p_1 = \sigma_v - 2c = 0$$

At point 2, (just above 2)

$$\sigma_v = r \cdot H = q$$
$$p_2 = rHK_a$$

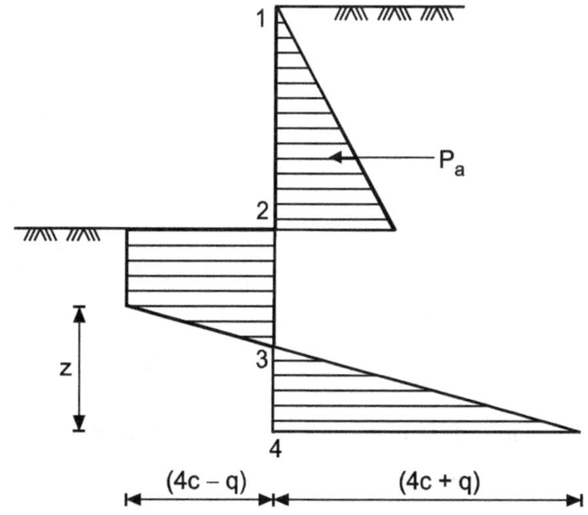

Fig. 5.32 : Pressure distribution diagram

At point 2,

Just below 2 to right, $p_{2r} = q - 2c$

Just below 2 to left, $p_{2l} = q + 2c = 2c$.

Net pressure at 2 to left of 2 = $p_{2l} - p_{2r} = 4c - q$

Net pressure at 4 = $p_{2r} - p_{2l} = 4c + q$

To determine the depth of embedment, we use the equilibrium equation.

$$\Sigma H = 0;$$

$$R_a + [(4c + q) + (4c - q)]\frac{\bar{z}}{2} = -(4c - q) D = 0$$

$$R_a (D + \bar{y}) - \frac{D^2}{2}(4c - q) + [(4c - q) + (4c + q)]\frac{\bar{z}}{2} \cdot \frac{\bar{z}}{3} = 0$$

Solving above equations; value of D can be computed and the computed value of 'D' is increased by 20% - 40%.

2. Anchored Sheet Piling:

- When the height of wall is more, cantilever sheet piles need penetration to considerable depth for stability. If sheet piles are anchored near their top, significant reduction in penetration depth can be made as it reduces the deflection near top.

- If anchored sheet piles are driven to a limited depth the deflection of beam is somewhat similar to that of vertical beam. Simply supported such sheet piles are called anchored piles with free earth supports.

- If anchored sheet piles are driven to greater depth the lower end of beam will behave just like partially fixed one. Such walls are called sheet pile wall with fixed earth support.

Fixed Earth Support :

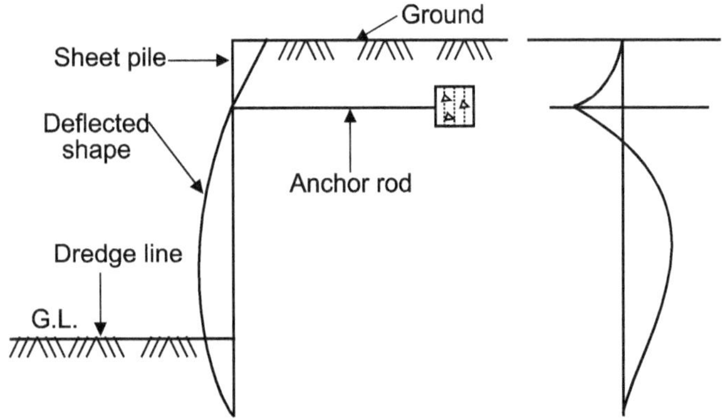

(a) Anchored sheet pile wall **(b) Bending moment diagram**

Fig. 5.33

Free Earth Support Method

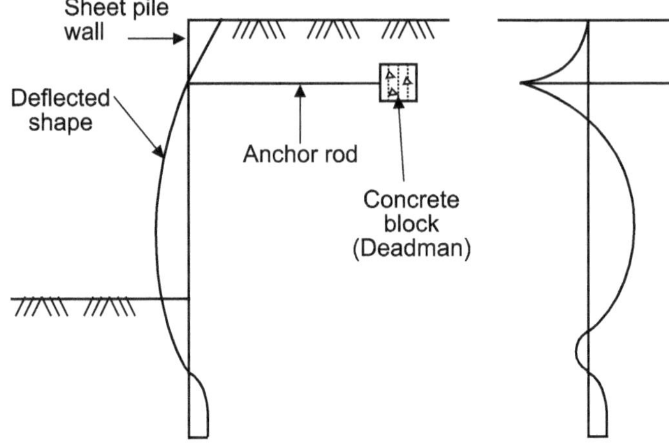

(a) Anchored sheet pile wall **(b) Bending moment diagram**

Fig. 5.34 : Fixed Earth Support Method

Design of anchored sheet pile wall (bulk head) consists of determining :

1. Depth of embedment.
2. Magnitude of tensile force.

1. Free Earth Support Method:

Method is based on following assumptions:

1. Sheet pile is perfectly rigid as compared to surrounding soil.
2. Sheet pile is free to rotate at anchor rod level with failure occurring due to rotation about anchor rod.
3. Active and passive earth pressure acting on sheet pile wall.

- ### Granular Soil:

Consider anchored sheet pile as shown in Fig. 5.35 with granular soil above and below dredge line.

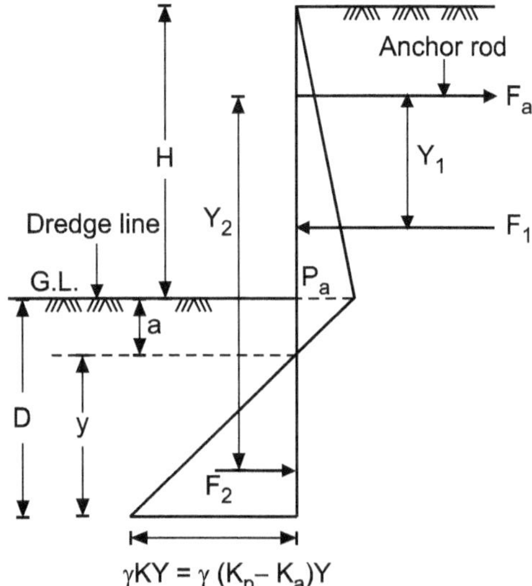

Fig. 5.35 : Pressure distribution diagram on anchored sheet pile wall

Let,

r : Unit weight of soil

Φ : Angle of internal friction

K_a, K_p : Coefficient of earth pressure

P_a : Intensity of active earth pressure at dredge line ($= rHK_a$)

F_a : Force in anchor rod

F_1 : Resultant of active earth pressure

F_2 : Resultant of passive earth pressure

a : Depth of point of zero pressure below dredgeline

At depth 'a' below dredgeline :

Active pressure $= P_a + r \cdot a \cdot K_a$

Passive pressure $= r \cdot a \cdot K_p$

$r \cdot a \cdot K_p = P_a + r \cdot a \cdot K_a$

$\therefore \quad a = \dfrac{P_a}{(K_p - K_a)}$

$\sum M = 0$ about anchor rod fi $\quad F_1 Y_1 = F_2 Y_2$

F_1 : Area of upper triangle

F_2 : Area of lower triangle

Y_1 : Distance of centre of gravity of upper triangle from anchor rod

Y_2 : Distance of centre of gravity of lower triangle from anchor rod

After substituting value of F_1, F_2, Y_1 and Y_2 in equation, we get an equation only in terms of 'y' which is solved for 'y'.

Depth of embedment,

$$D = y + a.$$

Value of 'D' is increased by 20-40% for safety purpose.

Force in anchor rod $= F_a = F_1 - F_2$

- **Granular Soil Above Dredgeline and Cohesive Soil Below Dredgeline :**

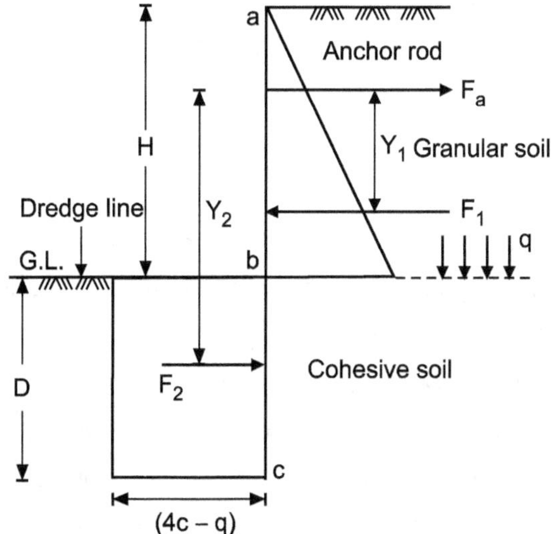

Fig. 5.36 : Distribution of earth pressure

Consider anchored sheet pile wall as shown in Fig. 5.36.

Earth pressure at $a = 0$

Earth pressure at $b = K_a \cdot r \cdot H = K_a \cdot q$

Active earth pressure just below 'b' $= q - 2c$

Passive earth pressure just below 'b' $= 2$

Net pressure at dredgeline $= 2c - (q - 2c) = 4c - q$

At any depth 'z' below dredgeline :

Active earth pressure $= (q + r \cdot z) - 2c$

Passive earth pressure $= rz + 2c$

Net pressure at depth z

$$= \text{Passive pressure} - \text{Active pressure}$$

$$= 4c - q$$

Applying $\sum M = 0$ about anchor rod,

$$F_1 Y_1 = F_2 Y_2$$

which is solved for 'D', calculated value of 'D' is increased by 20-40%.

Force in anchor rod is determined by applying $\sum H = 0$;

$$F_a = F_2 - F_1$$

5.10 COFFERDAMS

- Temporary structure generally constructed to prevent the water from entering in an area to facilitate construction projects in areas which are normally submerged such as bridges and piers. It is also constructed to remove soil from an area.

- Cofferdam is usually constructed large enough to provide adequate working space, (space for the proposed structure and working area for the workers around it). Once the area is enclosed by cofferdam the area is dewatered by pumping. Although cofferdam is constructed watertight yet during its life certain water will leak through foundation of cofferdam thus needs some pumping

Ideal Requirements of a Cofferdam

- It should be easy in construction at site of work.
- It should be stable against overturning.
- It should be watertight.
- It should occupy minimum possible area so as to provide sufficient space for containing permanent structures.
- It should provide least obstruction to the flow of water.
- It can be dismantled and reused easily.
- Its height should be more than HFL.
- Construction and maintenance cost should be low.

Uses of Cofferdam

Cofferdams are used in a variety of water environments, from small rivers to large harbors. Some of the more common uses for cofferdams include:

- **Hydroelectric Dam Construction:** Cofferdams are used to divert away from the shoreline of a river to allow for the foundations of a dam to be constructed. In this application, generally half of river width is enclosed by the cofferdam at a time to maintain overall flow.

- **Bridge Construction:** Cofferdams are used to divert water away from bridge foundation positions, either on the shore or within the waterway.

- **Ship Repair:** Sometimes cofferdams are used to generate a "Dry dock" conditions for a ship in order for repair to proceed. This generally occurs when the ship cannot be moved to an actual dry dock, and it can also be more cost effective in some cases.

- **Sunken Vessel Recovery:** Cofferdams can be used to expose sunken vessel in shallow waters to allow for recovery and repair if appropriate.

- **In Naval Architecture:** Cofferdam can also refer to the watertight bulkheads within a ship.

Types of Cofferdam

1. Dikes:
    - Earth dikes or Earth-fill cofferdam
    - Rock dikes or Rock-fill cofferdam
    - Sand bag dikes or Sandbag cofferdam.
2. Rock-fill crib cofferdam.
3. Single watt sheet pile cofferdam/single water timber sheet pile cofferdam.
4. Double wall cofferdam.
5. Braced cofferdam.
6. Cellular cofferdam.

Factors Affecting Selection of Type of Cofferdam

1. Extent of the area to be covered/enclosed.
2. Depth of water.
3. Possible wave-height.
4. Nature of velocity of flow (slow current/swift current).
5. Nature of bed on which cofferdam is to rest (pervious/impervious).

6. Scouring (bed erosion).
7. Local available material.
8. Availability of skilled labour.
9. Transport facility.
10. Risk of damage by floating debris or ice.

Suitability of Various Types of Cofferdam

1. **Earth-Fill Cofferdam** are suitable for low heads of water and if given surface protection, they can even suitable for half tide work, or on sites exposed to flowing water.

2. **Single Wall Sheet-pile Cofferdam** are suitable for restricted site area where cross bracings or ground anchors can be used.

3. **Double Wall Cofferdams or Cellular Sheet Piling** are used for wide excavation where self-supporting dams are required.

4. **Rock or Earth-Filled Timber Cribs** would be suitable for a remote site in undeveloped territory where heavy timber in log form is available and cost of importing and transporting steel sheet piling and its installation is comparatively very high.

The foundation soil condition is an important factor in the choice of cofferdam types. For example, sheet piling is not suitable for ground containing numerous boulders and cellular cofferdam could not be carried in deep deposits of clay.

5.10.1 Brief Description of Different Cofferdams

1. **Earthfill Cofferdam :**

- This is the simplest form of cofferdam and is suitable for sluggish rivers, lakes or other sheltered waters not subjected to high velocity of flow or wave action.

- These are usually restricted to low or moderate height of water (1.2 m – 1.5 m). It consists of an earth bank which is placed around the site to be enclosed as shown in Fig. 5.35.

- The thickness of bank should be sufficient to furnish required stability to cofferdam and to make it fairly watertight. Materials to be used for cofferdam depends on availability of material and site conditions usually a mixture of sand and clay used.

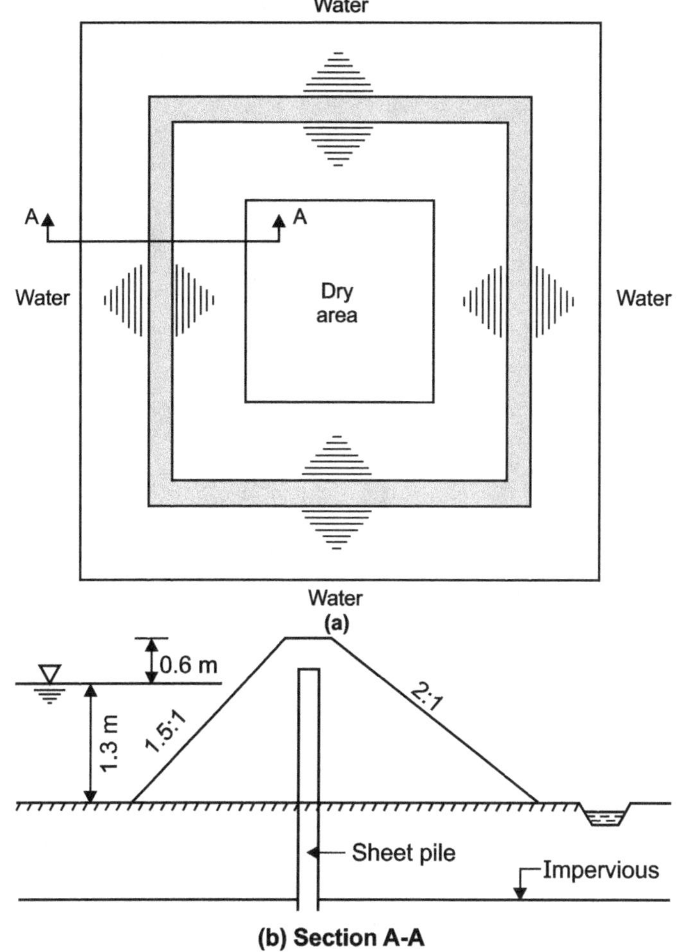

(a)

(b) Section A-A

Fig. 5.37 : Earthfill cofferdam

2. Rockfill Cofferdam:

- Rockfill cofferdams are similar in construction to earthfill cofferdam but due to inherent stability of material they can be formed with steeper slope than earthfill. They can be used for depth of water about 3 m and are suitable even in case of swift water.

- These cofferdams have the disadvantages of not being impervious. In case of low heights sufficient water tightness can be achieved by dumping impervious soil on outer face of cofferdam.

- Seepage will carry this material into the interstices of rock and fair degree of water tightness will gradually be attained. Rockfill cofferdam constructed properly can withstand over topping of water without any serious damage.

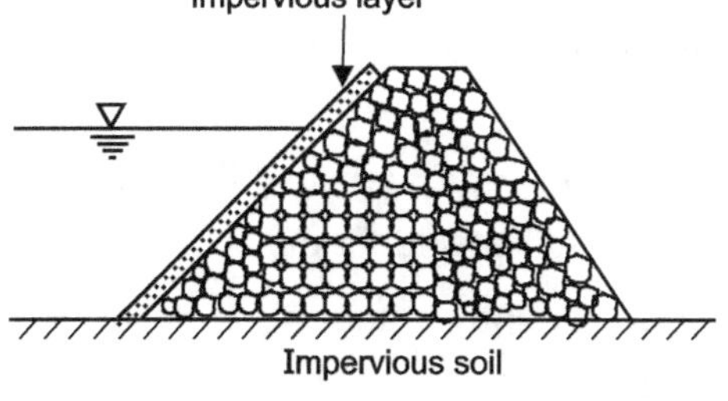

(a)

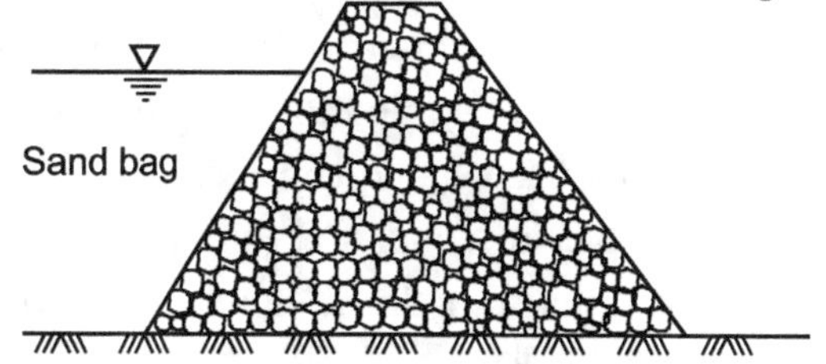

Fig. 5.38 : Rockfill cofferdam

3. Sand Bag Cofferdam :

- In this type of cofferdam, mixture of sand and clay is filled in bags and such bags are placed to form a cofferdam.

- Following points are noted :

  ➢ It is desirable to use empty bags as small quantity of cement present in empty bag will help in achieving watertightness.

  ➢ Bags should be partially filled. The bags can pack each other effectively and adjust with irregularities of soil at the bottom.

  ➢ Number of bags required should be carefully worked out with due allowance to wastage.

Fig. 5.39 : Sand-fill cofferdam

4. Rock-Fill Crib Cofferdam:

- Rock-fill crib cofferdam consists of timber cribs. A crib is a box or a cell open at the bottom and it essentially consists of a framework of horizontal timber laid in alternate courses.

- The pockets thus formed are then filled with rock or gravels or earth to give stability to crib against overturning and sliding.

- The timer to be used for construction of cribs may consist of rough logs or old sleepers from railway.

- This cofferdam is ideally suited for following situations:
  - Water current is swift.
  - There is danger of overtopping.
  - Depth of water is moderate.
  - Working space is limited.
  - Bed of stream is a hard rock.

5. Single Sheet-Pile Cofferdam:

- In this type of cofferdam there is a single row of cantilever sheet-piles the piles are sometimes heavily braced. Joints in the sheet-piles are properly sealed.

- Single sheet-pile cofferdams are generally used to enclose small foundation sites in water for bridging at shallow depth and is suitable for moderate velocity of flow of water and depth of water upto 4 m.

- Proportioning of wall are as below :

$$h \geq 0.25\,H \quad \text{...(for coarse sand and gravel)}$$
$$h \geq 0.5\,H \quad \text{...(for fine sand)}$$
$$h \geq 0.85\,H \quad \text{...(for silt.)}$$

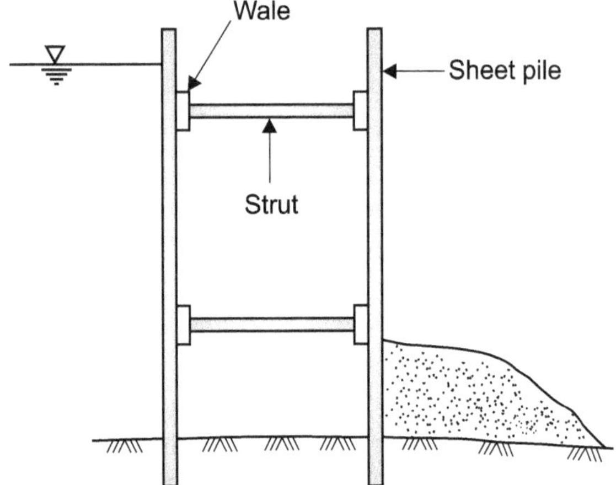

Fig. 5.40: Single sheetpile cofferdam

6. Double Sheetpile Cofferdam :

- Double sheetpile cofferdam consists of two straight vertical walls of sheeting tied to each other and space between them is filled with granular soil.

- Double sheetpilling cofferdam higher than 2.5 m should be strutted sometimes, inner berm is provided to increase its stability.

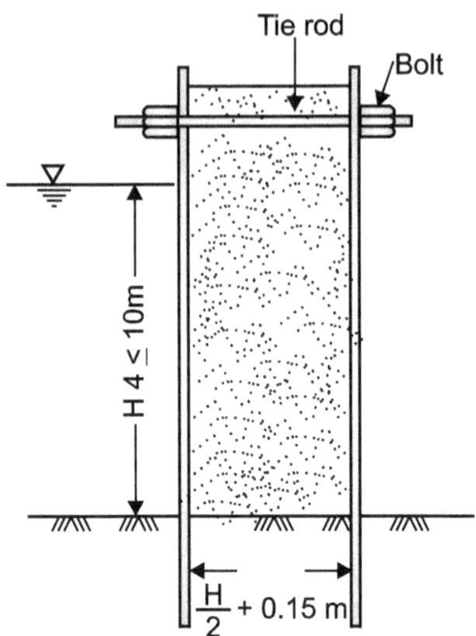

Fig. 5.41 : Double sheet pile cofferdam

7. Braced Cofferdam:

- When depth of excavation is greater than 10 m two rows of sheet-pile with interspace filled with murum is also inadequate/uneconomical. In such case braced cofferdam is used.

- This cofferdam is formed by driving vertical sheeting into the ground. Different sections of vertical sheeting are held in position by horizontal beams called wales. The wales in turn are supported by struts. These cofferdams are suitable upto a depth of 15 m.

Fig. 5.42 : Braced cofferdam

8. Cellular Cofferdams:

- The cofferdams consisting of cells are known as cellular cofferdams. These cofferdams are used as retaining structures. The retaining material is usually water whereas cell is filled with granular soil.

- These cells then interconnected enclose a working area in which construction operation or excavation operation can be carried out.

Functions of Cell Fill Material:

- It provides mass for stability of cofferdam.
- It reduces permeability of cofferdam.
- Ideal requirement of a fill material.
- It should have large coefficient of permeability so that free drainage is possible.
- It should have large angle of internal friction.
- It should contain very least amount of fine material.
- It should offer resistance against scouring due to turbulence of flow.
- It is practically impossible to get cell fill satisfying all these requirements.

Cellular cofferdam based on its shape and classified into following types:

1. Circular type cofferdam.
2. Diaphragm type cofferdam.
3. Cloverleaf type cofferdam.

Advantages :

- Cellular cofferdam is more economical when enclosed area and/or water head is large.
- More watertight.
- Cellular cofferdam is smaller than other types of cofferdam for given enclosed area and head thus obstruct less area of flow for navigation.
- Has more salvage value compared to other types.
- Relatively simple construction.

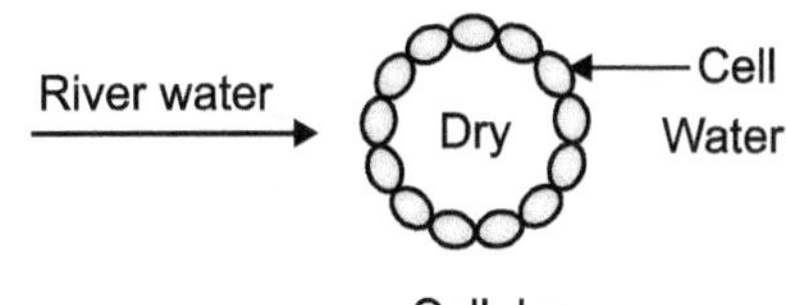

(a) Cellular cofferdam

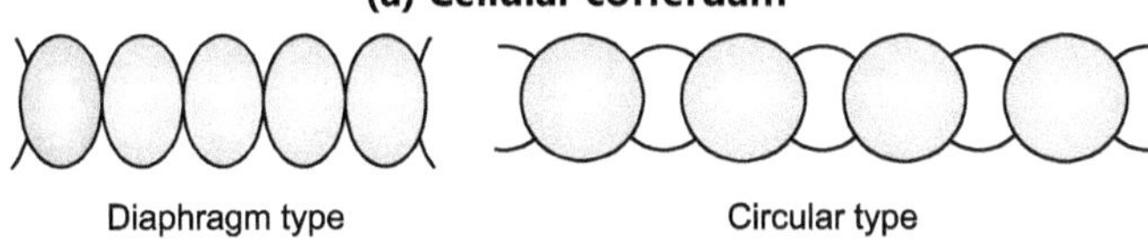

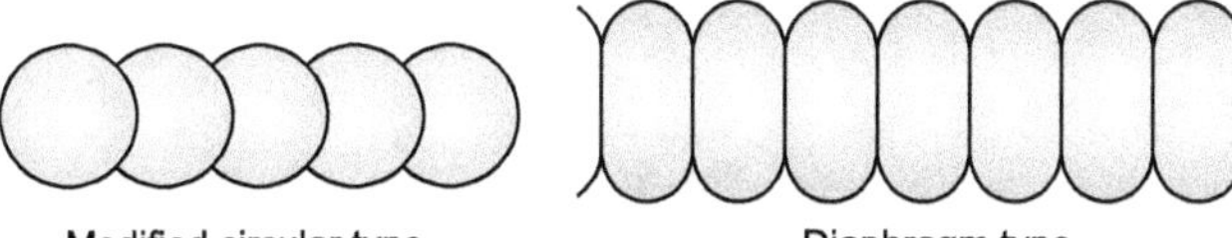

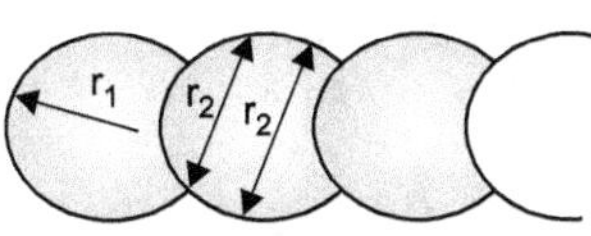

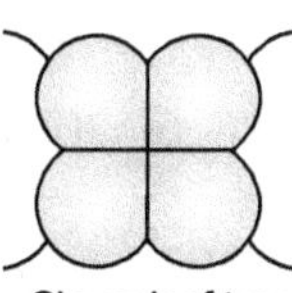

(b) Types of cellular cofferdam

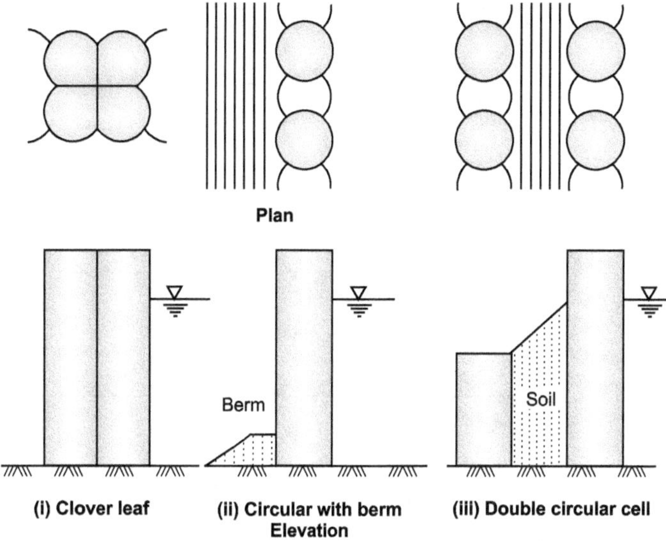

(c) Cellular arrangement to resist high lateral loads

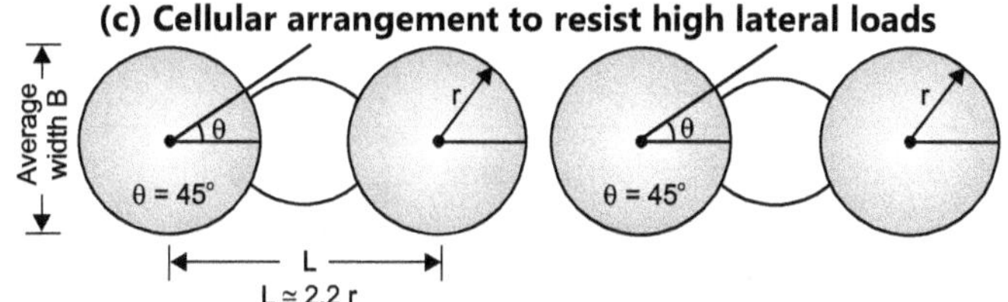

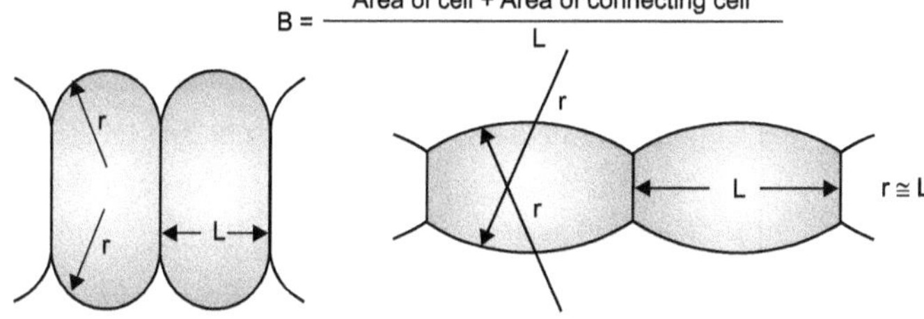

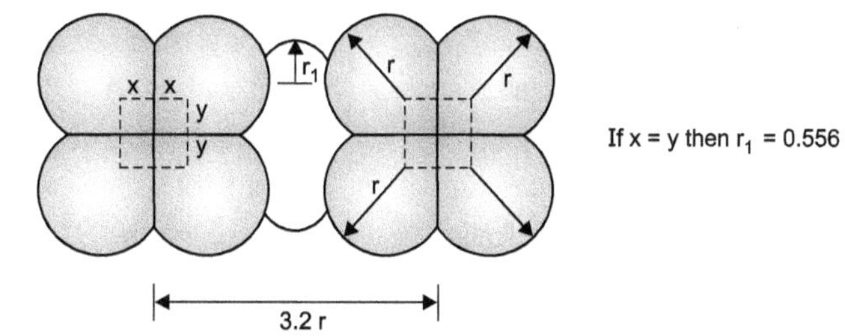

(d) Dimension of cellular cofferdams

Fig. 5.43

Table 5.8 : Merits and Demerits of Different Types of Cellular Cofferdams

		Merits	Demerits
Circular	1.	Each cell is self supporting and independent of adjacent cells thus failure of any one cell does not affect the other cell.	1. Diameter of circular cofferdam is limited by interlocking tension. Thus, cannot be used for high heads.
	2.	Can be used singularly or in a group or at end.	2. Great skill is required in setting and driving the pile.

...Conti.

	3. It can be filled with no regard to other cells.		
	4. Completed cells can be used as working platform for construction of adjacent cells.		
	5. Requires less number of piles per unit length of cofferdam.		
	6. Only this cofferdam can be constructed without protection in rough base with flowing water (v < 1.3 m/s)		
Diaphragm	1. Interlock tensions are smaller and uniform than those in circular type.	1. Cells are not independently stable.	
	2. It can be widened easily by increasing length of diaphragm (without affecting interlock tension) if required for stability.	2. Fill elevation in adjacent cells must be reasonably uniform to avoid distortion of walls.	
	3. They can be constructed cheaper and faster than circular cofferdam if done with floating equipment and progressive filling of cell.	3. Several templates are required during construction as each cell is not stable independently.	
Cloverleaf	1. Cells are independently stable.	1. Require more number of sheet piles.	
	2. Any size cell can be built to meet stability requirement but radii of arcs will be governed by interlock tension.	2. Fill elevation in adjacent compartments must be reasonably uniform to avoid distortion of walls of cofferdam.	
	3. It can be very effectively used as an end cells or corner cells.		

For Equal Section Modulus:

- Circular cofferdam is more economical for $r \leq 6$ m.
- Diaphragm cofferdam is more economical for $r > 6$ m.

For equal cross-sectional area diaphragm type is more economical than circular type.

SOLVED EXAMPLES

Example 5.1 : *Determine bearing capacity of pile by using the following data :*

(i) *Concrete pile = diameter 500 mm, length 8 m.*

(ii) *Angle of internal friction,* $\phi = 30°$.

(iii) *Bulk density of soil,* $\gamma = 9$ kN/m³.

(iv) *Cohesion 20 kN/m².*

(v) *Reduction factor,* $\alpha = 0.5$.

(vi) *Bearing capacity factor,* $N_c = 65$, $N_q = 35$, $N_\gamma = 18$.

(vii) *F.S. = 3.*

Solution : $\quad Q_u = Q_p + Q_s = q_u \cdot A_p + f_s \cdot A_s$

For Circular Pile (Terzaghi's equation) :

$$q_u = 1.3\, cN_c + \gamma_1 \cdot dN_q + 0.3\, \gamma_2\, BN_\gamma$$
$$= 1.3(20)\,(65) + 9(8)\,(35) + 0.3(9)\,(0.5)\,(18)$$
$$= 4234.3 \text{ kN}$$

$$A_p = \frac{\pi}{4}\,(0.5)^2 = 0.1963 \text{ m}^2$$

$$f_s = \left(\begin{array}{c}\text{Unit skin friction}\\ \text{due to cohesion 'c'}\end{array}\right) + \left(\begin{array}{c}\text{Unit skin friction}\\ \text{due to interlocking '}\phi\text{'}\end{array}\right)$$

$$= \alpha c + K \cdot \bar{q} \cdot \tan \delta$$

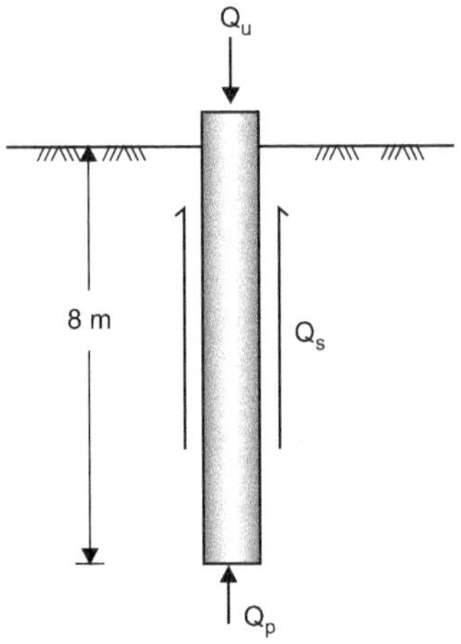

Fig. 5.44

Assuming $K = 1.5$, $\delta = \dfrac{\phi}{2} = 15°$

$$f_s = (0.5)\,(20) + 1.5\left(\frac{1}{2} \times 9 \times 8\right) \tan 15°$$

$$= 24.47 \text{ kN/m}^2$$

$$A_s = \pi(0.5)\,(8.0)$$

$$= 12.57 \text{ m}^2$$

$$Q_u = (4234.4)(0.1963) + (24.47)(12.57)$$

$$= 1138.7 \text{ kN}$$

$$\text{Safe load} = \frac{Q_u}{3} = 379.56 \text{ kN} \cong 380 \text{ kN}$$

Example 5.2 : *Determine whether failure is by group or individual action, using the following data :*

  (i) No. of piles in group = 16.

  (ii) Diameter of pile = 45 cm.

  (iii) Spacing both ways = 1.5 m c/c

  (iv) Cohesion = 50 kN/m²

  (v) Neglect bearing

  (vi) Length of pile = 10 m

  (vii) Shear mobilization factor for each pile = 0.7.

Solution : For Individual Action :

$$Q_u = 16 \times \text{Capacity of single pile}$$

$$= 16 \left[q_p \, A_p + f_s \cdot A_s \right]$$

Neglecting bearing ($q_p = 0$)

$\therefore \qquad Q_u = 16 \, f_s \cdot A_s$

$$= 16 \times (0.7 \times 50) \times \pi (0.45)(10)$$

$$= 7916.81 \text{ kN} \qquad \qquad \dots \text{(i)}$$

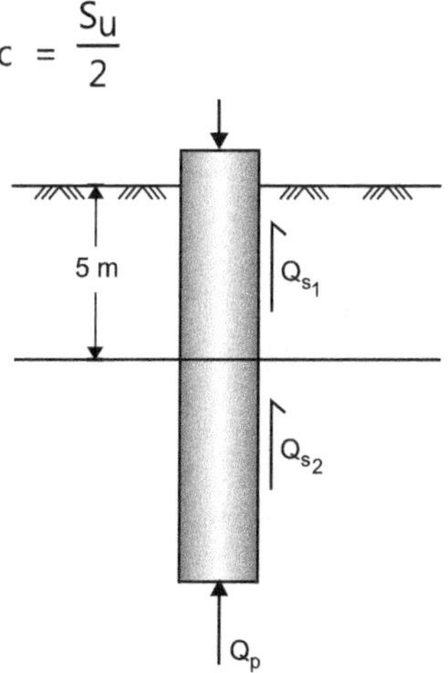

Fig. 5.45

For Block Action :

$$A_g = (4.95)(4.95) = 24.5025$$

$$P_g = 4(4.95) = 19.8 \text{ m}$$

$$Q_u = q_p \cdot A_g + f_s \cdot P_g \cdot L$$

Neglecting bearing ($q_p = 0$)

$$Q_u = (0.7)(50)(19.8)(10)$$

$$= 6930 \text{ kN} \qquad \qquad \dots \text{(ii)}$$

As capacity of group by block action is less than that by individual action [comparing (i) and (ii)]. Thus, failure is by group action.

Example 5.3 : *Decide the type of failure for groups of piles, use following data :*

  (i) No. of pilesin a group - 16.

  (ii) Diameter of pile - 45 cm.

  (iii) Length - 10 m.

  (iv) Center to Center distance of piles - 1.5 m.

  (v) c = 50 kN/m².

  (vi) Shear mobilization factor - 0.7.

Solution : By Block Action : Neglecting weight of soil block

$$Q_g = 0.7 \times 50 \times [2 \, (3 \times 1.5 + 0.45) + 2 \, (3 \times 1.5$$

$$+ \, 0.45)] \times 10 + 9 \times 50 \times (4.95)^2$$

$$= 17956 \text{ kN}$$

By Individual Action :

$$Q_I = 16 \left[0.7 \times 50 \times \pi \times 0.45 \times 10 + 9 \times 50 \times \frac{\pi}{4} (0.45)^2 \right]$$

$$= 9061.92 \text{ kN}$$

Group fail by individual action as $Q_I < Q_g$.

Example 5.4 : *40 cm diameter pile is to carry a load of 380 kN. The pile is passing through two clayey layers.*

  (i) Layer is of 5 m thickness and has unconfined compressive strength of 90 kN/m².

  (ii) Layer has unconfined compressive strength of 200 kN/m².

Determine the length of pile to carry the load with a F.S. = 2. Assume coefficient of adhesion as 0.5.

Solution : Let the length of pile = L

$$c = \frac{S_u}{2}$$

Fig. 5.46

For (I) layer, $\quad c_1 = \dfrac{90}{2} = 45 \text{ kN/m}^2$

For (II) layer, $\quad c_2 = \dfrac{200}{2} = 100 \text{ kN/m}^2$

$$Q_u = f_s \cdot A_s + q \cdot A_p$$
$$= \alpha \cdot c_1 \cdot A_{s_1} + \alpha c_2 \cdot A_{s_2} + g \cdot c_2 \cdot A_p$$
$$= Q_{s_1} + Q_{s_2} + Q_p$$
$$Q_{s_1} = (0.5)(45)(\pi \times 0.4 \times 5) = 141.37 \text{ kN}$$
$$Q_{s_2} = (0.5)(100)(\pi \times 0.4 \times (L - S)]$$
$$= 62.83 (L - S)$$
$$Q_p = 9 \times 100 \times \frac{\pi}{4}(0.4)^2 = 113.1 \text{ kN}$$
$$Q_u = 141.37 + 62.85 (L - S) + 113.1$$
$$(380)(2) = 62.85 L - 58.78 \Rightarrow L = 13.04 \text{ m}$$

Provide L = 13.1 m

Example 5.5 : *Determine static bearing capacity of a pile. Use following data.*

Diameter of pile = 300 mm.

Length of pile = 8 m.

Embeded in (i) layer of 3 m thick with γ_b = 10 kN/m³,

$\phi = 30°$, $c = 10 kN/m^2$,

and (ii) layer of 8 m thick with γ_b = 9 kN/m³. $\phi = 0$, $c = 60 kN/m^2$.

Use $\alpha = 0.5$ for (II) layer.

Use $k = 0.5$ for (I) layer and neglect cohesion.

Also work out % contribution by each component in total bearing capacity.

Solution : $Q_u = Q_s + Q_p$
$$Q_s = Q_{s_1} + Q_{s_2}$$

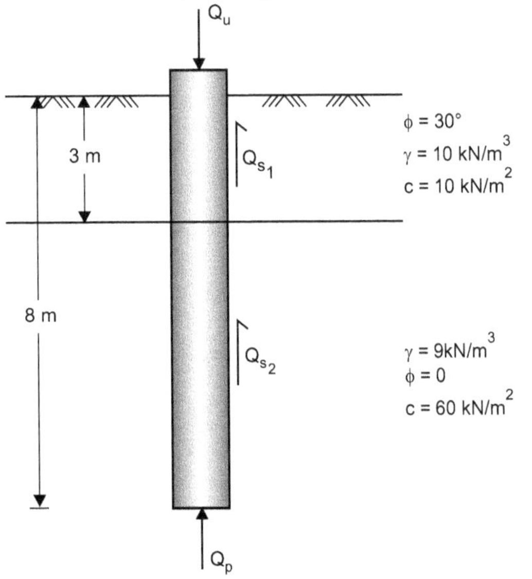

Fig. 5.47

$$Q_{s_1} = [k \cdot \bar{q} \cdot \tan \delta] \left[\text{As assuming } \delta = \frac{\phi}{2} \right]$$
$$= 0.5 \left[\frac{1}{2}(10)(3) \right] [\tan 15°] = 2.0 \, A_s$$

$$Q_{s_1} = 2 \times \pi \times 0.3 \times 3 = 5.65 \text{ kN}$$
$$Q_{s_2} = \alpha \cdot c \cdot (\pi \times 0.3 \times 5)$$
$$= 0.5 \times 60 \times \pi \times 0.3 \times 5 = 141.37 \text{ kN}$$
$$Q_s = 141.37 + 5.65 = 147.02 \text{ kN}$$
$$Q_p = q \cdot A_p = 9 \times 60 \times \frac{\pi}{4}(0.3)^2 = 38.17 \text{ kN}$$
$$Q_u = 147.02 + 38.17 = 185.19 \text{ kN}$$

Contribution by bearing $= \dfrac{38.17}{185.19} \times 100 = 20.6\%$

Contribution by skin friction = 79.4%.

Example 5.6 : *20 number of 30 cm diameter piles are arranged in 5 rows and 4 columns. The piles pass through clayey strata. Determine suitable spacing so as to achieve maximum group efficiency. Assume block failure.*

Solution : $S = \dfrac{m \cdot n \cdot \pi \cdot d}{2(m + n - 2)} = \dfrac{5 \times 4 \times \pi \times 0.3}{2(5 + 4 - 2)}$

$$= 1.34 \text{ m c/c}$$

Example 5.7 : *50 cm diameter pile passing through a clayey strata having uniform unconfined compressive strength of 56 kN/m² is to carry a load of 400 kN. Assuming F.S. = 2.5, determine the length of pile to carry the load. Assume co-efficient of adhesion as 0.95.*

Solution : Let L - Length of pile.

$$Q_u = Q_s + Q_p \qquad \ldots \text{(i)}$$

Here $Q_s = f_s \cdot A_s = \alpha \cdot c \cdot A_s$
$$= (0.95)(28)(\pi \times 0.5 \times L) = 41.78 \, L$$

and $Q_p = q_u \cdot A_p = 9c \cdot A_p$
$$= 9 \times 28 \times \frac{\pi}{4}(0.5)^2 = 49.48$$

$\therefore \quad Q_u = 41.78 \, L + 49.48$

$\therefore (400)(2.5) = 41.78 \, L + 49.48$

$\therefore \qquad L = 22.75 \text{ m}$

$\therefore \qquad L \cong 23.0 \text{ m}$

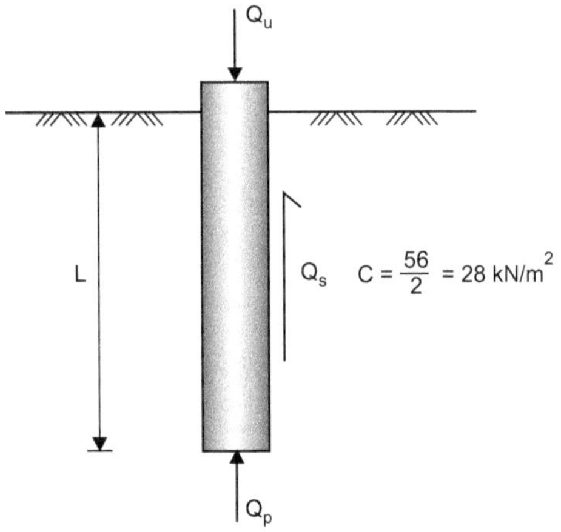

Fig. 5.48

Example 5.8 : *Group of piles consisting of m rows and n columns is passing through clayey strata. If each pile is of uniform diameter 'd' then prove that for maximum group. "η" spacing 'S' should be*

$$S = \frac{m \cdot n \cdot \pi \cdot d}{2\,(m + n - 2)}$$

Solution : 'η' will be maximum when group capacity by block action and individual action is same.

Capacity of block action

$$= f_s \cdot A_s + q_p \cdot A_p$$

$$= \alpha \cdot c \cdot P_g \cdot L + gc \cdot B_g \cdot L_g \qquad \dots (i)$$

Capacity by individual action

$$= N \cdot \left[\alpha \cdot c \cdot \pi\, dL + gc \cdot \frac{\pi}{4} d^2 \right] \qquad \dots (ii)$$

Contribution by bearing is very less in comparison with skin friction. So it is neglected.

$$\alpha \cdot c \cdot P_g \cdot L = N \cdot \alpha \cdot c \cdot \pi dL$$

$$\therefore \alpha \cdot c \cdot [2\,(B_g + L_g)]\,L = m \cdot n \cdot \alpha \cdot c \cdot \pi dL$$

Put $\quad B_g = (m - 1)\,s + d$

and $\quad L_g = (n - 1)\,s + d$

$$\therefore 2\,[(m - 1)\,s + d + (n - 1)\,s + d] = m \cdot n \cdot \pi \cdot d$$

$$\therefore 2\,[(m + n - 2)\,s + 2d] = mn\pi d$$

$$s = \frac{mn\pi d - 4d}{2\,(m + n - 2)} = \frac{m \cdot n \cdot \pi \cdot d}{2\,(m + n - 2)}$$

Example 5.9 : *Find out group capacity of piles by the four methods.*

  (i)   Feld's rule   (ii)   Converse labarre

  (iii) Block failure   (iv) Individual

Use following data :

- *15 piles, 300 mm diameter.*
- *Depth of piles = 8 m*
- *Strata with c = 25 kN/m²*
- *Spacing of pile = 0.8 m c/c*
- *$\alpha = 1.0$*
- *Unit weight of soil = 10 kN/m³*

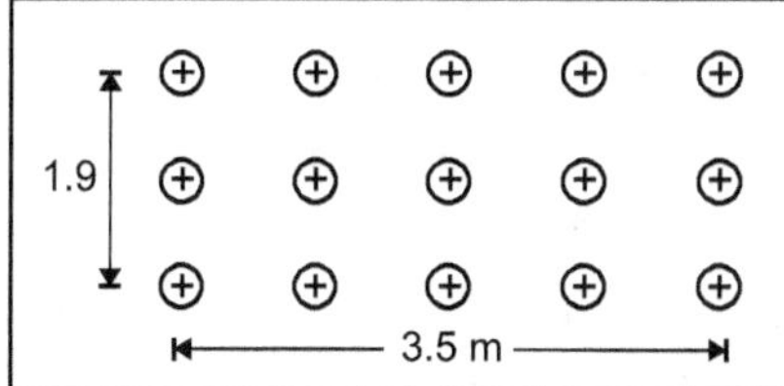

Fig. 5.49

Solution : By Individual Action :

$$Q = 15 \times \left[1.0 \times 25 \times \pi\,(0.3)\,(8) + 9\,(25)\left(\frac{\pi}{4}\right)(0.3)^2 \right]$$

$$= 15 \times 204.4 = 3066 \text{ kN}$$

By Block Action :

$$Q = [1.0 \times 25 \times (10.8) \times 8 + 9 \times 25 \times 3.5$$
$$\times 1.9] - 10 \times 3.5 \times 1.9 \times 8$$
$$= 3124.25 \text{ kN}$$

By Feld's Rule :

$$\gamma = \frac{4\,(1 - 3/16) + 8\,(1 - 5/16) + 3\,(1 - 8/16)}{15}$$

$$= \frac{4\,(13) + 8\,(11) + 3\,(8)}{15 \times 16} = 0.68$$

$$Q_g = \eta \cdot NQ = 0.68 \times 3066 = 2095 \text{ kN}$$

By Converse Labarre :

$$\eta = 1 - \frac{\theta}{90} \left[\frac{m(n - 1) + n(m - 1)}{m \cdot n} \right]$$

$$\left\{ \because \theta = \tan^{-1}\left(\frac{0.3}{0.8}\right) = 20.55° \right\}$$

$$= 1 - \frac{20.55}{90} \left[\frac{3(5 - 1) + 5(3 - 1)}{3 \times 5} \right] = 0.66$$

$$Q_g = (0.66)\,(3066)$$

$$= 2023.56 \text{ kN}$$

Example 5.10 : *In a two layered cohesive soil, bored piles of 400 mm are installed. The top layer has a thickness of 5 m and bottom one is of considerable depth. Unconfined compressive strength of top clay layer is 45 kN/m² and that of bottom is 100 kN/m². Determine the length of bored pile required to carry a safe load of 380 kN, allowing factor of safety of 2.0. Assume $\alpha = 0.5$, $N_c = 9.0$.*

Solution : Let x - length of pile in second layer

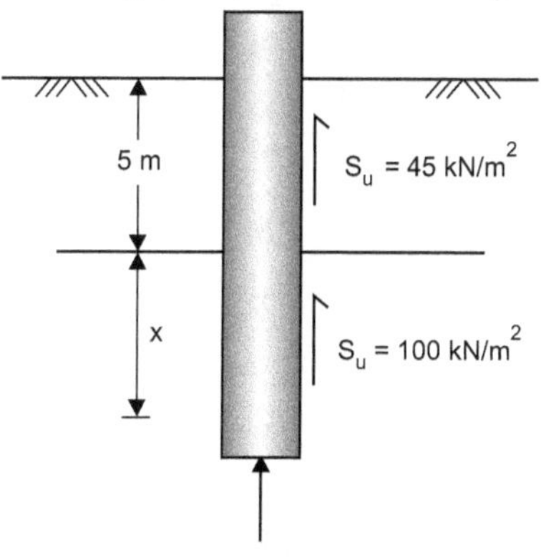

Fig. 5.50

$$c_1 = \frac{S_u}{2} = 22.5 \text{ kN/m}^2$$

$$c_2 = 50 \text{ kN/m}^2$$

$$Q_u = Q_{s_1} + Q_{s_2} + Q_p$$

$$= \alpha c_1\,(\pi d L_1) + \alpha c_2\,(\pi d L_2) + q_p \cdot \frac{\pi}{4}\,d^2$$

$\therefore (380)\,(2.0) = (0.5)\,(22.5)\,(\pi \times 0.4 \times 5) + (0.5)\,(50)$

$$(\pi \times 0.4 \times x) + 9 \times 50 \times \frac{\pi}{4}\,(0.4)^2$$

$$760 = 70.68 + 31.41x + 56.55$$

$\therefore \qquad x = 20.14 \text{ m}$

Total length of bored pile required $= 20.14 + 5 = 25.14 \cong 25.2$ m

---

Example 5.11 : *A pile 300 mm diameter 8 m deep is installed in a stratum having shearing resistance angle of 30°. A cohesion for this stratum may be taken as 10 kN/m². Value of adhesion factor is 0.8 and density of stratum is 1800 kg/m³. Find ultimate capacity of pile.*

Solution : Ultimate capacity of pile is given by :

$$Q_u = f_s A_s + q_p A_p$$

Neglecting bearing

$$Q_u = f_s A_s$$

f_s = Unit skin friction of soil

$$= \alpha c + K \bar{q} \tan\delta$$

$$= (0.8)\,(10) + 1.5 \left(\frac{1}{2}\,(17.66 \times 8)\right)\tan 15^{\circ}$$

$$= 36.39$$

$$Q_u = (36.39)\,(\pi \times 0.3 \times 8)$$

$$= 274.38 \text{ kN/m}^2$$

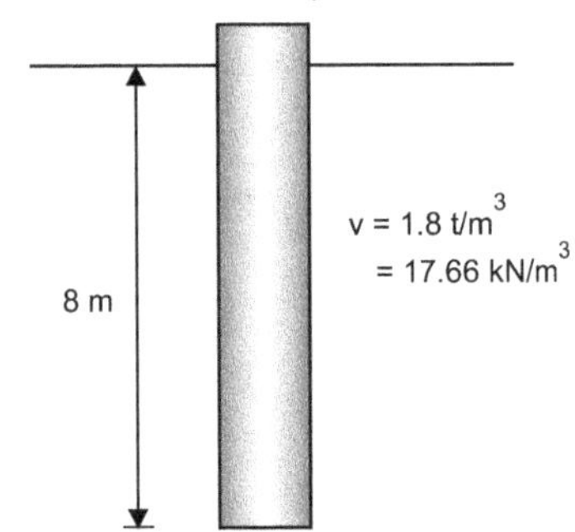

Fig. 5.51

Example 5.12 : *Compute the 'η' of pile group consisting of 20 piles which are arranged in four rows, if diameter of pile is 400 mm and spacing is 1m center to center by using.*

(a) Converse Labbare's formula

(b) Seiler Keeney's formula

(c) Feld's rule

Solution : Fig. 5.45 shows plan of group

$$m = \text{no. of rows} = 05$$

$$n = \text{no. of columns} = 05$$

$$\theta = \tan^{-1}\left(\frac{0.4}{1.0}\right) = 21.8^{\circ}$$

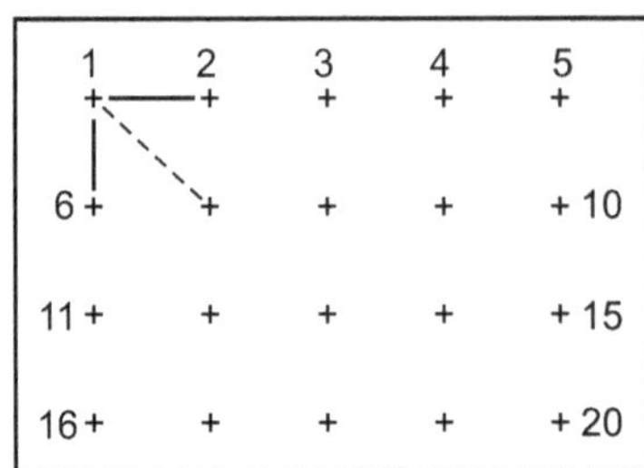

Fig. 5.52

Converse Labbare's formula :

$$\eta_9 = 1 - \frac{\theta}{90}\left[\frac{m(n-1) + n(m-1)}{m \cdot n}\right]$$

$$= 1 - \frac{21.8}{90}\left[\frac{4 \times 4 + 5 \times 3}{20}\right] = 62.45\%$$

Seiler Keeney's formula :

$$\eta_9 = \left[1 - 0.479\left(\frac{S}{S^2 - 0.093}\right)\left(\frac{m + n - 2}{m + n - 1}\right)\right] + \frac{0.3}{m + n}$$

$$= \left[1 - 0.479\left(\frac{1}{1 - 0.093}\right)\left(\frac{7}{8}\right)\right] + \frac{0.3}{9}$$

$$= 57.12\%$$

Feld's rule :

All corner piles have three adjacent piles, nearby value of all such piles

$$= 1 - 3 \times \frac{1}{16} = 0.8125.$$

Piles 2, 3, 4, 10, 15, 17, 18 and 19 have five adjacent piles, value of such piles

$$= 1 - 5 \times \frac{1}{6} = 0.6875$$

Piles 7, 8, 9, 12, 13 and 14 have eight adjacent piles, value of such piles

$$= 1 - 8 \times \frac{1}{16} = 0.50$$

$$\eta = \frac{2(0.8125) + 8(0.6875) + 6(0.50)}{16}$$

$$= 63.28\%$$

Example 5.13 : *A single acting steam hammer weighing 2500 kg and falling through a height of 2.2 m drives a pile to an average penetration of 0.88 cm under the last few blows. What will be the allowable load of the pipe ? Use engineering News formula.*

Solution : Load carrying capacity of pile is given by,

Engineering News formula,

$$Q_n = \frac{WH}{6\,(S + 0.25)} = \frac{2500 \times 220}{6\,(0.88 + 0.25)}$$

$$= \frac{550000}{6.78} = 81.48 \times 10^3 \text{ kg}$$

Example 5.14 : *A wooden pile is driven by drop hammer weighing 2000 kg and having free fall of 5 m. The final settlement is 1.5 cm. The allowable load, using engineering News formula will be how much ?*

Solution :
$$Q_n = \frac{WH}{6\,(S + 2.5)} = \frac{2000 \times 500}{6\,(1.5 + 2.5)}$$
$$= \frac{1 \times 10^6}{24} = 41.66 \times 10^3 \text{ kg}$$

Example 5.15 : *A pile of 0.35 m × 0.35 m cross-sectional area penetrated to soft soil having C = 0.75 kg/cm² for a length of 15 m and finally rests on hard soil. Calculate load carrying capacity by skin friction.*

Solution : Load carrying capacity for a square pile,
$$Q = 4\,d_a\,L \times C_a = 4 \times 0.35 \times 15 \times \alpha c$$

Assuming $\alpha = 1$ for soft clay

∴ $\quad Q = 4 \times 0.35 \times 15 \times 7.5 = 157.5$ tonnes

Example 5.16 : *A circular pile section with 0.35 m diameter and length 10 m penetrates a deposit of clay having C = 5 kN/m² and mobilizing factor α = 0.8. Calculate load carrying capacity by skin friction.*

Solution :
$$Q = \pi d_a\,L \times C_a$$
∴ $\quad Q = \pi \times 0.35 \times 10 \times C_a$
$$C_a = \alpha \times C$$
$$= 0.8 \times 5 = 4$$
∴ $\quad Q = \pi \times 0.35 \times 10 \times 4$
$$= 43.98 \text{ tonnes}$$

Example 5.17 : *The load carrying capacity of an individual friction pile is 200 kN. What is the total load carrying capacity of a group of 8 such piles with group efficiency factor 0.8 ?*

Solution : Total load carrying capacity of group $= 200 \times 8 \times 0.8 = 1280$ kN

Example 5.18 : *Skin friction capacities of a 40 cm diameter driven concrete pile for the portions A, B and C are 17 kN, 63 kN and 503 kN respectively and the point load capacity is 12000 kN/m². What will be total point load capacity ?*

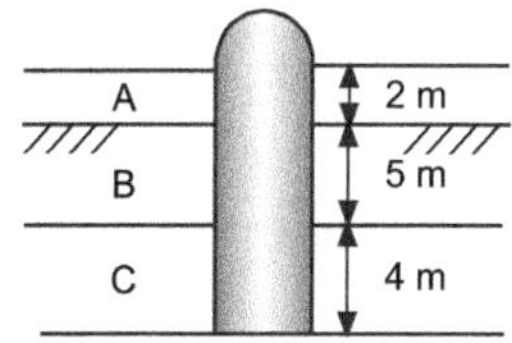

Fig. 5.53

Solution : Load carrying capacity by skin friction,
$$= \alpha \times C \times A_p = \alpha \times C$$
$$= 0.8 \times 5 \times 4 \times \frac{\pi}{4} \times 0.4^2 \times 10$$
$$= 20.10 \text{ kN}$$

Example 5.19 : *A group of 16 piles of 50 cm diameter is arranged with a center to center spacing of 1.0m. The piles are 9m long and are embedded in soft clay with cohesion 30 kN/m². Bearing resistance may be neglected for the piles- Adhesion factor is 0.6. Determine the ultimate load capacity of the pile group.*

Solution : Individual action, we get,
$$Q_g = N\,[fs \cdot As] = 16\,[0.6 \times 30 \times \pi\,(0.5)g]$$
$$Q_g = 4069.44 \text{ kN}$$

Group action, $Q_g = fs \cdot As$

Width of pile group,
$$B = 3 \times 1 + 0.5 = 3.5 \text{ m}$$
$$As = 4 \times 3.5 \times 9 = 126 \text{ m}^2$$
$$fs = C = 30 \text{ kN/m}^2$$
$$Q_g = 126 \times 30 = 3780 \text{ kN} < Qg$$

Governing Group action,
$$Q_g = 3780 \text{ kN}$$

Example 5.20 : *Design square pile group to carry 500 kN load in clay with an unconfined compressive strength of 80 kN/m². The pile are 30cm diameter and 8 m long and adhesion factor 0.6. Use FS =3. .*

Solution :
$$Q_u = Q_s + Q_p$$
Here, $\quad Q_s = f_s \cdot A_s = \propto \cdot c \cdot A_s$
$$= 0.6 \times 40 \times (\pi \times 0.3 \times 8) = 180.96 \text{ kN}$$
and, $\quad Q_p = q_u \cdot A_p = 9_c \cdot A_p = 9 \times 40 \times \frac{\pi}{4} \times (0.3)^2$
$$= 25.44 \text{ kN}$$
$$Q_u = 206.40 \text{ kN}$$

We get,
$$\text{Safe Load} = \frac{Qu}{3} = \frac{206.40}{3} = 68.8 \text{ kN}$$

∴ No. of piles $= \dfrac{500}{68.80} = 7.26$ say 7.

We provide 9 piles for square pattern,
∴ $\quad Q_g = 9 \times 68.80 = 6.19.2 \text{ kN} > 500 \text{ kN}$

Example 5.21 : *Determine the capacity of pile by using following data. Diameter of pile= 600 mm, length = 7m, φ = 30°, soil density = 17 kN/m³. c = 20 kN/m², reduction factor, α =0.5, Nc = 65, Nq = 35, Nγ = 18, factor of safety = 3.*

Solution : We have,
$$Q_u = Q_p + Q_s = q_u \cdot A_p + f_s \cdot A_s$$
For Circular Pile (Terzaghi's Equation) :
$$q_u = 1.3\,cN_c + \gamma_1 \cdot dN_q + 0.3\,\gamma_2\,BN_\gamma$$
$$= 1.3\,(20)\,(65) + 17 \times 7 \times 35 + 0.3 \times 17 \times 0.5 \times 18$$
$$= 5900.9 \text{ kN}$$

$$A_p = \frac{\pi}{4}(0.5)^2 = 0.1963 \text{ m}^3$$

So we get,

$$f_s = \left(\begin{array}{c}\text{unit skin friction}\\ \text{due to cohesion 'c'}\end{array}\right) + \left(\begin{array}{c}\text{unit skin friction}\\ \text{due to interlocking '}\phi\text{'}\end{array}\right)$$

$$= \propto_c + k \cdot \bar{q} \, \tan \delta$$

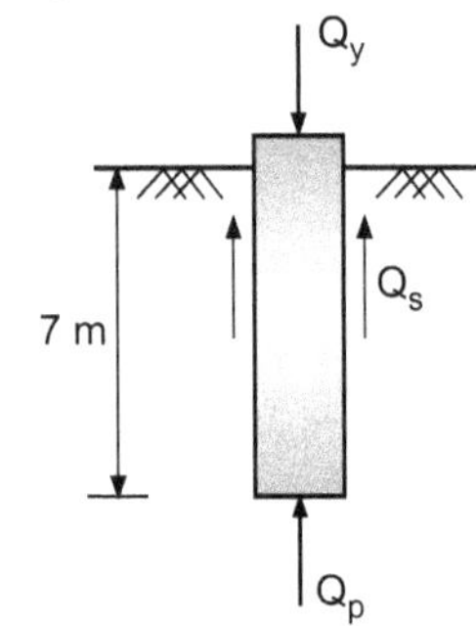

Fig. 5.54

Assuming,　　$k = 1.5, \delta = \dfrac{\phi}{2} = 15°$

Now,　　$f_s = (0.5) \times 20 + \left[1.5\left(\dfrac{1}{2} \times 17 \times 7\right)\right] \tan 15°$

$$= 33.91 \text{ kN/m}^2$$

$$A_s = \pi(0.5)(7.0) = 10.99 \text{ m}^2$$

$$Q_u = (5900.9)(0.1963) + (33.91)(10.99)$$

$$= 1531.02 \text{ kN}$$

$$\text{safe load} = \frac{Q_u}{3} = 510.34 \text{ kN} \approx 510 \text{ kN}$$

Example 5.22 : *A group of piles consists of 15 piles arranged in three rows and five columns. Compute the efficiency of pile group by Feld's rule.*

Solution : By Feld's rule,

$$\text{We get,} \quad \gamma = \frac{4\left(1 - \dfrac{3}{16}\right) + 8\left(1 - \dfrac{5}{16}\right) + 3\left(1 - \dfrac{8}{16}\right)}{15}$$

$$= \frac{4(13) + 8(11) + 3(8)}{15 \times 16} = 0.69$$

EXERCISE

1. It is proposed to conduct a conventional pile load test. For this purpose suggest with sketches on following points :
   (a) Detailed layout in section with naming component.
   (b) Procedure in ten steps.
   (c) Observations, plotting and interpretation of test results as per BIS.
2. Write short note on rigid block method of failure of friction piles.

3. Discuss the concept of negative skin friction in piles with a sketch. How would you determine magnitude of the same in cohesive, non-cohesive and cohesionlesssoils ?
4. Write detailed notes on :
   (i) Pile test by cyclic load method (procedure).
   (ii) Pile test by cyclic load method (interpretation).
5. Gravity type pile load test is to be conducted. Suggest the following :
   (i) Schematic arrangement in plan and section and name component parts.
   (ii) Procedure in 10 steps.
   (iii) Observation and preservation of results.
   (iv) Interpretation of results.
6. Answer the following with sketches :
   (i) Rigid block method of failure of group of piles.
   (ii) Negative skin friction in piles.
7. Explain how do you decide bearing capacity of single pile by any one of the following method :
   (i) Static method.
   (ii) Dynamic method.
   (iii) Load test method.
8. Explain with sketches and in ten steps the construction of bore and cast in-situ piles.
9. It is proposed to conduct pile load test by (gravity load) as per details given below. For this proposal answer the following with sketches.
   (i) Sketch of test in plan and in section, naming component part and indicate their function.
   (ii) Method of applying load and measurement of settlements.
   (iii) Recording and plotting of observation.
   (iv) Assessment of safe load as per BIS.
10. Explain the following with sketches.
    (i) Tree type exhaustive classification of piles with basis of classification.
    (ii) Safe load carrying capacity of pile by static method and C, ϕ soils and pile resting on hard strata.
11. State and explain with sketches static formula for determining bearing capacity of single vertical pile subjected to vertical load under the following condition.
    (i) Piles in 'C' soil.
    (iii) Piles in 'C-ϕ' soil.
    (ii) Piles in 'ϕ' soil.
    (iv) Piles resting on hard strata through organic soils.
12. What is caisson ? Describe different types of caissons.
13. Write short note : Shapes and factors deciding shape of caisson foundation.

14. Draw three simple sketches and point out the use of open caisson, box caisson and pneumatic caisson.

15. Compare in a tabular form box caisson, open caisson and pneumatic caisson w.r.t. following points :

    (a) Sketch in plan and section.

    (b) Component parts.

    (c) Method of sinking.

    (d) Load bearing capacity.

    (e) Specific use with illustrations.

    (f) Method of construction.

16. Discuss the tabular form with sketches open caisson, box caisson and pneumatic caisson w.r.t. purpose, component parts, method of sinking for each of these methods.

17. Explain the procedure of caisson sinking using 'Sand Island Method'.

18. Draw neat sketch of pneumatic caissons indicating all the components/parts. Explain the precautionary measures adopted during-construction of a pneumatic caissons. Give conditions favouring pneumatic caisson as a choice of foundation systems.

19. Draw a neat sketch of pneumatic caisson, label various parts and explain its working.

20. Explain with neat sketch if needed

    (a) Box caisson.

    (b) Caisson disease.

21. Draw three simple sketches and point out use of open caissons, box caissons and pneumatic caissons.

22. Write advantages and disadvantages of floating caissons as compared to open caisson.

23. What is caisson sickness ? How is it controlled ?

24. Draw a neat sketch of various components of caisson foundations, name components and parts.

25. Explain sinking of pneumatic caissons. What is the function of 'air-lock' in the pneumatic caissons.

26. What is caisson disease ?

27. Enlist five important components of well foundation.

28. Draw neat sketches of components of double D shaped well in plan and section and explain the functions of various components with sketches.

29. Write short notes : Various parts of well foundation.

30. A well foundation is to be constructed on dry sandy bed to a depth of 10 m or 50 m. Starting with beginning explain with sketches how would you proceed and complete the job.

31. A circular well of 8 m external diameter and steining thickness of 1 m is to be sunk in position. For this proposal explain the following with sketches.

    (a) Casing of well on sandisland for initial height of 5 m and its sinking.

    (b) Equipments used.

    (c) Precautions for sinking.

    (d) Difficulties met.

    (e) Rectification of any one difficulty.

32. A railway bridge pier with double 'D' shaped well foundations has following details.

    (a) Top of pier about RL 100.00 m with steel girder resting on it.

    (b) Pier dimension in plan 2 m × 10 m.

    (c) Top of well cap slab 90 m RL with 2 m depth.

    (d) Double D shaped well foundation of 5 m × 11 m in plan with staining at 1 m thick and digging wells of minimum 3 m × 4 m in plan.

    (e) Bottom of well resting on rock at RL 77.00 m and properly anchored.

    (f) HWL = 97 m.

    (g) LWL = 92 m.

    (h) GL = 88 m.

    For this proposal.

    (i) Draw neat sketches in plan and section and name all components and parts.

33. Explain with neat sketch sand island method for well sinking.

34. In connection with a circular well explain following with sketches :

    (a) Sand island method.

    (b) Method of sinking.

    (c) Rectification of tilted well.

35. Explain with neat sketches difficulties met and remedial measures adopted in sinking of well.

36. Explain with sketches arrangements involved in three different types of sheet piles and circumstances under which each is used.

37. Draw three different sketches of anchored sheet piles and show on them elastic deflection, pressure distribution and moment diagram.

38. Compare with sketches empirical pressure distribution diagrams on braced cofferdam in dense sand and compact clay which would produce more earth pressure for same depth.

39. Explain the terms with sketches:
    (i) Free earth support.
    (ii) Fixed earth support in connection with anchored sheet piles for their B.M.

40. State and explain five different methods of anchorage of sheet piles.

41. Draw the sketches of structural arrangements involved in :
    (i) Cantilever sheet pile.
    (ii) Anchored sheet pile.
    (iii) Braced sheet pile.
    (iv) Cellular (diaphragm) sheet piles. Name component parts.

42. Explain with sketch single cantilever sheet pile, anchored sheet pile, two independent sheet piles with internal supports and double wall sheet piles and state how do they get strength in progressive manner.

43. A cantilever sheet pile has following details:
    (i) Cantilever depth ... 'h'.
    (ii) Depth of embarkment ... 'd'.
    (iii) K_a, K_p coefficients of active and passive earth pressures.

    Show by simple method that the depth of embedment is : $d \times \sqrt[3]{K_p/K_a} - h$.

44. Discuss with sketches the construction of under reamed piles and the equipments used for this purpose.

PROBLEMS FOR PRACTICE

1. Determine static bearing capacity of a pile. Use following data:
   (a) Diameter of pile - 300 mm.
   (b) Length – 8 m
   (c) Embedded in Layer I of 3 m thick with γ_b = 10.0 kN/m^3, ϕ = 30°, c = 10 kN/m^2 and Layer II of 8 m thick with γ_b = 9.0 kN/m^3, ϕ = 0, c = 60 kN/m^2
   (d) Use α for Layer II as 0.5
   (e) Use K = 0.5 for Layer I and neglect cohesion.
   Also work out % contribution by each component in total bearing capacity.
   [**Ans. :** 185.15 kN, 79.38% by skin friction]

2. 50 cm diameter pile passing through a Clayey strata having uniform unconfined compressive strength of 56 kN/m^2 is to carry a load of 400 kN. Assuming F.S. = 2.5, determine the length of pile to carry the load, assume coefficient of adhesion as 0.95.

[**Ans. :** 22.75 m]

3. A group of 16 numbers of 40 cm diameter piles is to carry to load of 2500 kN. The substrata consists of clay having unconfined compressive strength of 48 kN/m^2 and effective unit weight of 9 kN/m^3. Determine the length pile if F.S. = 3 against shear failure is required. Assume adhesion coefficient = 0.9.
   [**Ans. :** 17.21 m]

4. A group of piles consists of 15 piles arranged in three rows and five columns. Compute efficiency of pile group by following method. Assume diameter of piles = 300 mm spacing 0.75 m centre to centre.
   1. Converse – Labbare's formula
   2. Los-Angles formula
   3. Feld's rule.
   [**Ans. :** 64.47%, 71.72%, 68.33%]

5. Calculate depth of embedment of sheet pile by using approximate method, use following data :
   (a) Projected length of sheet pile – 6 m.
   (b) Water table – 3 m from top.
   (c) Properties of soil upto 3 m depth – γ = 19 kN/m^3, ϕ = 30°.
   (d) Properties of soil below 3 m depth – γ' = 10 kN/m^3, ϕ = 30°.
   Draw a neat sketch of proposal and solution may be in full numerical.
   (**Ans.** D = 9.74 m)

6. For anchored sheet pile calculate the distance of zero earth pressure location by using following data :
   (a) G.L. = 100.00 m R.L.
   (b) Length of pile = 10 m from ground.
   (c) Dredge line = 94 m R.L.
   (d) ϕ of back fill = 30°.
   (e) Unit weight, γ = 18 kN/m^3, γ_{sat} = 20 kN/m^3.
   (f) Water table about R.L. 97.00 m.
   Draw neat sketch of proposal.
   (**Ans.** 1.04 m below dredge line)

7. A cantilever sheet pile has following details:
   (a) Length of pile = 9 m.
   (b) Depth of embedment = 5 m.
   (c) Earth fill, ϕ = 30°.
   Work out factor of safety for this. If this factor is to be reduced to 1.0, find out condition.
   (**Ans.** F.S. = 1.54, Reduce depth of embedment = 4.33 m)

STABILITY OF SLOPE

6.1 INTRODUCTION

- Slope is the surface of soil which is inclined at some angle with horizontal usually between $0 - 70^0$. (When angle is more than 70^0 then it is called wall). Slopes are required whenever soil is to be supported at two different levels without retaining wall in between them.

- Slopes are commonly required for following structures Highway, railway, canal, earth dam, trench etc. Cost of the earthwork would be minimum if the slopes are made steepest. However steep slopes may not be stable.

- A compromise has to be made between economy and safety i.e. slopes provided are neither too steep nor too flat (steepest slope which are stable and safe should be provided)

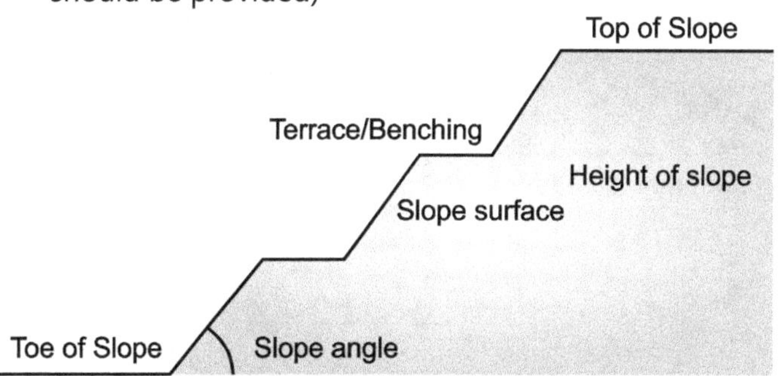

Fig. 6.1 : Slope nomenclature

Slope Classification: Slopes are classified based on various criteria's which are as given below :

- Based on agents of formation
  1. Natural
  2. Artificial or man made
- Based on extent of slope
  1. Infinite
  2. Finite
- Based on composition of slope
  1. Homogeneous
  2. Heterogeneous or Non homogeneous

Concept of Stability of Slope:

- Whenever there is difference of level between two surfaces of soil gravitational force (weight) try to cause movement from the higher to lower level unless the shearing resistance within the material is sufficient to withstand these forces.

- The forces which tend to cause slippage are known as actuating forces (Gravitational, seepage, earthquake etc.) and the force which oppose slippage (shear strength) are called resisting forces.

- The concept of stability analysis of slope is based on the fact that on every plane or curved surface across the slope shear resistance must be more than the shearing stress due to actuating forces; when this condition is not satisfied slope will fail.

- Slope stability is arguably the most complex and challenging of all the sub-disciplines of geotechnical engineering, and is often the least understood.

- Generally, failure occurs due to natural or man-made causes. Natural failures primarily occur because of stresses imposed by weight of the soil mass itself and by changing soil properties.

- Man-made failures occur when the slope is physically altered. Irrespective of the mechanism causing failure, a slope fails when the imposed stresses exceed the shear strength of the soil along the failure.

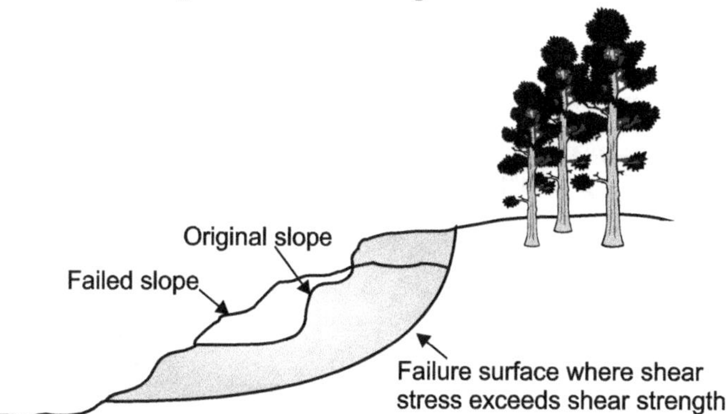

Fig. 6.2 : Typical slope failure

Assumptions in Analysis of Slope:

Actual analysis of slope failure is complex and indeterminate to simplify analysis following assumptions were made in slope analysis

1. Problem is 2 – D.
2. Shear strength of the soil is known and can be represented by Coulomb's law.
3. Seepage condition and water level are known and the corresponding pore-pressure can be estimated.
4. Shear strain at all points along critical surface are large enough to mobilize all available shear strength.

6.2 CAUSES OF SLOPE FAILURE

- Generally, failure occurs due to natural or man-made causes. Natural failures primarily occur because of stresses imposed by weight of the soil mass itself and by changing soil properties. Man-made failures occur when the slope is physically altered. Irrespective of the mechanism causing failure, **a slope fails when the imposed stresses exceed the shear strength of the soil along the failure.**

- Failures of natural and man-made slopes are generally attributable to any activity that results in either an increase in soil stress or a decrease in soil strength. The specific causes of slope instability are varied and depend on the nature of the soil, pore water pressure, climate, and stress within the soil mass (static and dynamic).

- Specific examples that cause a net increase in stresses include an increase in the unit weight of the soil through rainfall, loads imposed by fills or structures at the top of a slope or excavation at the toe of a slope, movement of water levels (such as rapid drawdown in a reservoir), earthquakes, and water pressure in cracks within the slope. These are shown in Fig. 6.3.

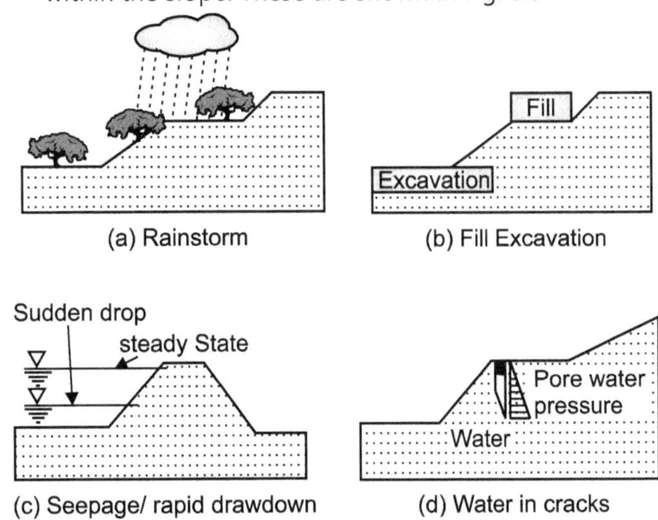

Fig. 6.3 : Various causes of slope failure

- Water plays a role in many of the processes that reduce strength, and water is also involved in many types of loads on slopes that increase shear stresses. Therefore, virtually every slope failure involves the destabilizing effects of water in some way, and often in more than one way.

- Another factor involved in most slope failures is the presence of soils that contain clay minerals. The behavior of clayey soils is much more complicated than the behavior of gravels, sands, and non-plastic silts, which consist of chemically inert particles.

- The mechanical behavior of clays is affected by the physicochemical interaction between clayparticles, the water that fills the voids between the particles, and the ions in the water.

- When a slope fails, it is usually not possible to pinpoint a single cause that acted alone and resulted in instability. For example, water influences the stability of slopes in so many ways that it is frequently impossible to isolate one effect of water and identify it as the single cause of failure

6.2.1 Types of Slope Failure

1. Translational failure
2. Rotational failure
3. Wedge failure
4. Compound failure
5. Miscellaneous failure

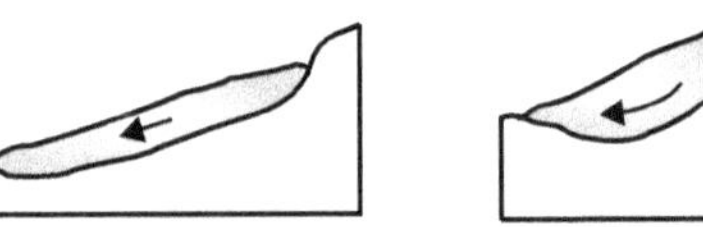
Translational Failure　　**Rotational failure**　　**Wedge failure**

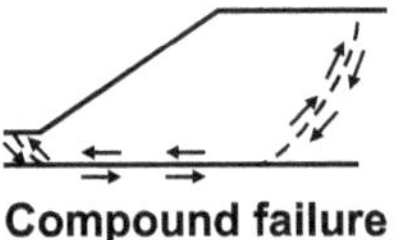
Compound failure

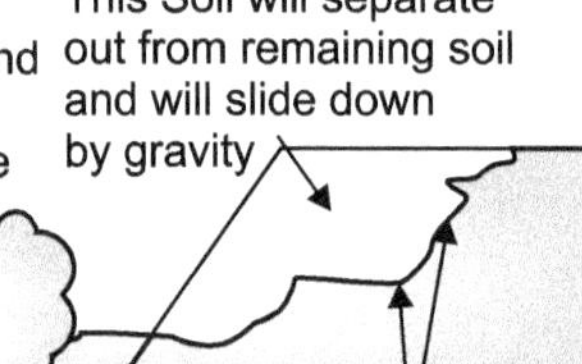

Miscellaneous failure

Fig. 6.4 : Types of slope failure

Face failure
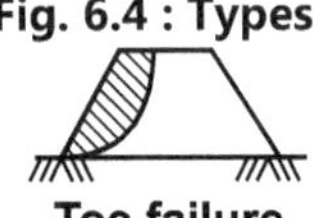
Toe failure
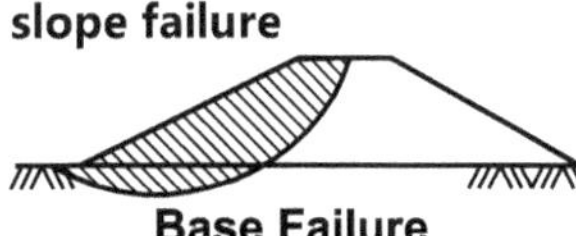
Base Failure

Fig. 6.5 : Types of rotational failure

- ➢ If failure surface intersects the slope above toe then it is called **face failure**,
- ➢ If at the toe then it is called **toe failure** and
- ➢ If it intersects below toe then it is called **base failure**.

6.3 FACTOR OF SAFETY

It is dimensionless number which will ensure the safety of a structure and it depends on :

- The method of stability analysis used.
- The method used to determine the shear strength.
- The degree of confidence in the reliability of subsurface data.
- The consequences of a failure.
- How critical the application is.

A minimum factor of safety as low as 1.25 is used for highway embankment side slopes. This value of the safety factor should be increased to a minimum of 1.30 to 1.50 for slopes whose failure would cause significant damage such as end slopes beneath bridge abutments, major retaining structures and major roadways such as regional routes, interstates, etc.

6.3.1 Types of Factor of Safety

1. F.S. with respect to shear strength.
2. F.S. with respect to cohesion.
3. F.S. with respect to friction.
4. F.S. with respect to moment.
5. F.S with respect to height.

These factor of safety are defined as below :

1. F.S. with respect to shear strength

$$F_s = \frac{\text{shear strength of soil}}{\text{mobilised shear stress}}$$

2. F.S. with respect to cohesion

$$F_s = \frac{\text{cohesive strength of soil}}{\text{mobilised shear stress}}$$

3. F.S. with respect to friction

$$F_s = \frac{\tan \varphi}{\tan \varphi_m} \approx \frac{\varphi}{\varphi_m}$$

4. F.S. with respect to moment

$$F_s = \frac{\text{overturning moment}}{\text{resisting moment}}$$

5. F.S with respect to height

$$F_s = \frac{\text{critical height}}{\text{actual height}}$$

6.4 ANALYSIS OF SLOPE

Slope stability analysis is performed to assess the potential for failure of the slope by rupture. The primary objective of a stability analysis is to determine the factor of safety (FS) of a particular slope, to predict when failure is imminent, and to assess remedial treatments when necessary. In many practical situations, an analytical assessment of stability can be made

6.4.1 Analysis of Infinite Slope

1. Dry Soil:

- Consider an infinite slope having slope angle β as shown in figure. Consider small element of this slope of width b; various forces acting on this are as shown.
- For an infinite slope, the forces on the two ends of the block will be identical in magnitude, opposite in direction, and collinear.
- Thus, the forces on the ends of the block exactly balance each other and can be ignored in the equilibrium equations.

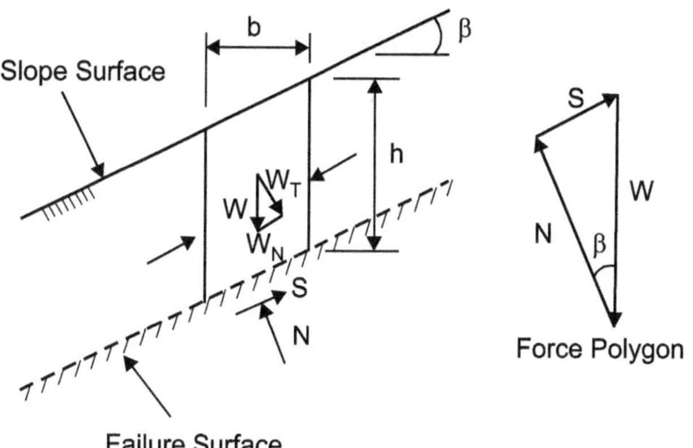

Fig. 6.6: Failure of infinite slope

- Summing forces in directions perpendicular and parallel to the slip plane gives the following expressions for the shear force, (S), and normal force, (N), on the plane:

$$S = W \cos \beta \ \text{ and } \ N = W \sin \beta$$

But $\quad W = \gamma bh = $ weight of element of soil

$$\text{Normal stress} = \frac{N}{\text{Area parallel to failure surface}}$$

$$= \frac{\gamma bh \cos \beta}{b/\cos \beta} = \gamma h \cos^2 \beta$$

$$\text{Shear stress} = \frac{S}{\text{Area parallel to failure surface}}$$

$$= \frac{\gamma bh \sin \beta}{b/\cos \beta} = \gamma h \sin \beta \cos \beta$$

$$\text{Factor of safety} = \frac{\text{Shear strength}}{\text{Shear stress}} = \frac{c + \sigma \tan \varphi}{\gamma h \cos^2 \beta}$$

$$= \frac{C + \gamma h \cos^2 \beta \tan \varphi}{\gamma h \sin \beta \cos \beta}$$

When $c = 0$, we get $\quad FS = \dfrac{\tan \varphi}{\tan \beta}$

Height of the slope at which slope is on the verge of failure i.e. shear stress on failure surface is equal to shear strength of soil (FOS = 1)

Substituting FOS = 1 and h = HC in above equation we get

$$C + \gamma H_c \cos^2 \beta \tan \varphi = \gamma H_c \sin \beta \cos \beta$$

$$\gamma H_c \sin \beta \cos \beta - \gamma H_c \cos^2 \beta \tan \phi = C$$

$$\gamma H_c \cos^2 \beta (\tan \beta - \tan \varphi) = C$$

$$\frac{C}{\gamma H_C} = \cos^2 \beta (\tan \beta - \tan \varphi)$$

Term $\dfrac{c}{\gamma H_c}$ is called stability number.

$$H_c = \frac{c}{\gamma \cos^2 \beta (\gamma \tan \beta - \tan \varphi)}$$

- For submerged slope

$$H_c = \frac{C}{\gamma' \cos^2 \beta (\tan \beta - \tan \varphi)}$$

- Slope with seepage parallel to slope

$$H_c = \frac{C}{\cos^2 \beta \,(\gamma \tan \beta - \gamma' \tan \varphi)}$$

2. When the Seepage is Parallel to Slope :

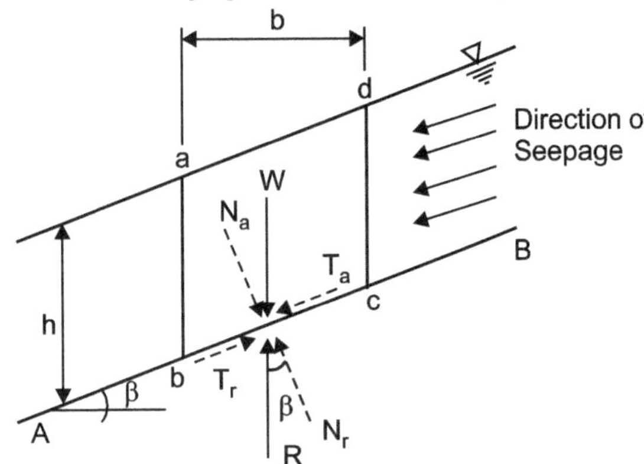

Fig. 6.7 : Infinite slope with seepage parallel to slope

$$FOS = \frac{C}{\gamma_{sat}\, h \cos^2\beta \tan \beta} + \frac{\gamma'}{\gamma_{sat}} \frac{\tan \varphi}{\tan \beta} \qquad \text{...for c-}\varphi \text{ soil}$$

for granular soil $C = 0$

$$\therefore \qquad FOS = \frac{\gamma'}{\gamma_{sat}} \frac{\tan \varphi}{\tan \beta}$$

General Case :

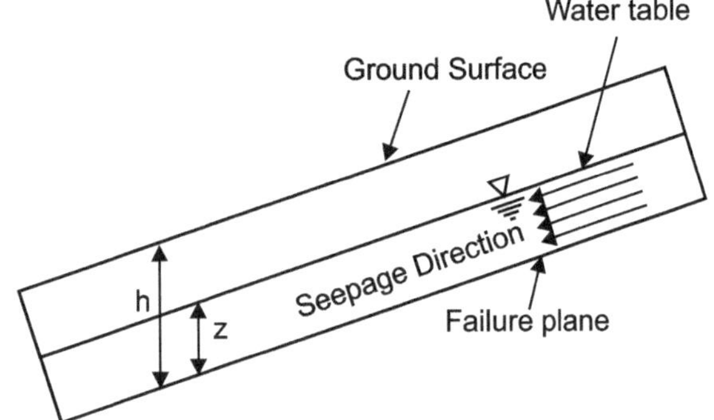

Fig. 6.8 : Infinite slope with seepage parallel to slope (partially submerged)

$$FOS = \frac{C + h \cos^2\beta \,[(h - z)\,\gamma + z\gamma']\tan \varphi}{h \sin \beta \cos \beta \,[(h - z)\,\gamma + Z\gamma_{sat}]}$$

$$FOS = \frac{c + h \cos^2\beta \,[(1 - m)\,\gamma + m\gamma']\tan \phi}{h \sin \beta \cos \beta \,[(1 - m)\,\gamma + m\gamma_{sat}]}$$

Where   c and φ : shear parameters of soil

        γ : bulk unit weight of soil above water table

      γ_{sat}, and γ' : Saturated and submerged bulk unit weight of soil

        β : Slope angle

        h : Height of slope

        m : normalized height of water table above failure plane = z / h

        Z : Height of water table above failure surface

        h : Height of slope

6.4.2 Analysis of Finite Slope

The finite slope can be analyzed by any of following method :

1. Swedish circle method or slip circle method or method of slice.
2. Friction circle method.
3. Taylors method.

1. Swedish Circle Method:

Investigation carried out in Sweden on slope failures indicated that the failure surfaces resembles the arc of a circle. Fellenious developed a method for stability analysis of a slope assuming a circular failure surface, which is known as the Swedish circle method.

- **Stability Analysis for Cohesive Soil ($\varphi = 0$ analysis)**

Fig. 6.9 shows finite slope whose stability is to be determined. A trial slip surface of radius r is assumed and factor of safety of the slope is determined for the assumed trial surface. For analysis purpose unit length of slope is considered.

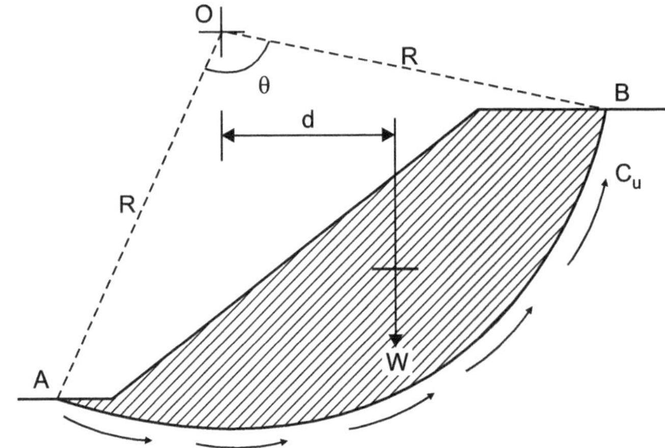

Fig. 6.9 : Finite slope in purely cohesive soil

Let   R : radius of slip circle

         c : cohesion of soil (undrained)

         O : center of slip circle

         W : weight of sliding soil (soil above failure surface)

         d : Horizontal distance between center of circle and the point where weight acts (lever arm)

         θ : central angle

         $\hat{L}$: Arc length

Moment causing sliding of soil mass downward = disturbing moment = MD = Wd

Moment resisting sliding tendency of soil mass = resisting moment = $MR = c\hat{L}R$

Factor of safety = Resisting moment / disturbing moment

$$FS = \frac{c\hat{L}R}{Wd} = \frac{cR^2\theta}{Wd}$$

Analysis is repeated for number of trial slip surfaces and the FOS is determined in eachcase. The slip circle corresponding to minimum FOS is the critical slip circle.

Note:

- Position of centroid of sliding wedge can be determined by dividing wedge of soil into small slices.

- If soil is non-homogeneous or slope has benches, then it should be divided into small elemental areas and then moment of all such areas need to be taken about center of rotation.

C – φ Analysis (Method of Slices):

- For soil which has both cohesion and friction component of shear strength, that is, c – φ soil, the shear strength along the slip surface is also contributed by the frictional component, which is a function of normal stress.

- The normal stress varies at every point on the slip surface both in magnitude and direction hence total sliding soil mass is divided into number of slices (vertical). The force between the slice is neglected and each soil is assumed to act independently as a column of soil.

- The weight of each slice is assumed to act at its Centre. Weight of each slice is resolved into normal (N) and tangential component (T). The normal component passes through the center of slip circle and does not cause any moment. However, the tangential component will cause a driving moment (= Ti × r) where r is radius of slip circle. For some slices tangential component may cause resisting moment in which case it should be considered negative.

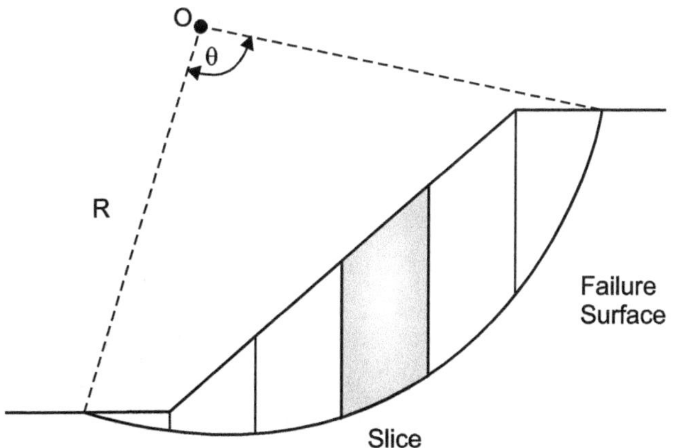

(a) Sliding soil mass divided in to number of slices

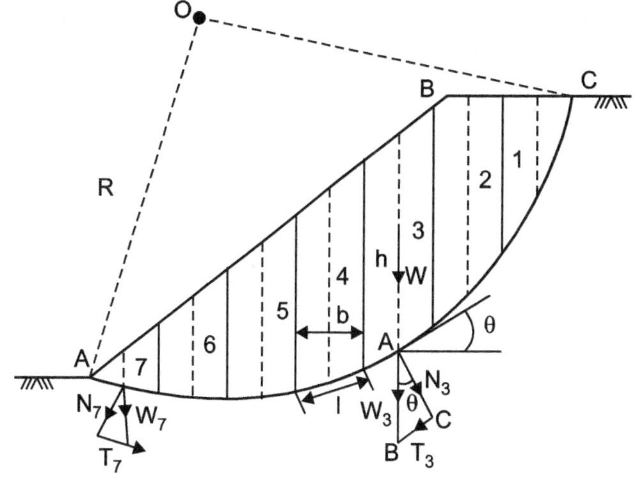

(b) Component of weight of slice in to normal and tangential component

Fig. 6.10

$$FOS = \frac{c\hat{L} + \tan\varphi \sum N}{\sum T}$$

When soil is submerged we need to consider pore-water pressure (U) in that case factor of safety is given by

$$FOS = \frac{c\hat{L} + \tan\varphi \sum (N - U)}{\sum T}$$

Where
$\sum N$ – algebraic sum of normal component of the weight of slice

$\sum T$ – algebraic sum of tangential component of the weight of slice

$\sum U$ – algebraic sum of pore pressure

c – cohesion of soil,

φ – angle of internal friction

$\hat{L}$ – arc length of failure surface = $r\theta$

θ – central angle

Calculations will be done in tabular column.

Table 6.1

Slice No.	Width of Slice	Area of Slice	Weight of slice	θ	Component of Weight	
					Normal = W cos θ	Tangential = W sin θ
					ΣN =	ΣT =

Note: If the calculation of factor of safety is done partially by graphical construction and partially by analytical method then it is called semi-analytical method, and if it is done entirely by graphical construction then it is called graphical method.

- **Fellenious Method of Locating Center of Slip Circle:**

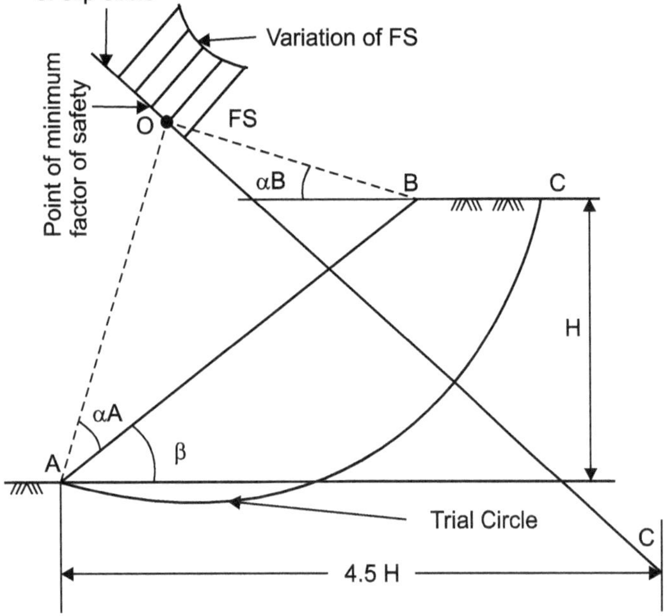

Fig. 6.11

Table 6.2

Slope	Slope Angle	Direction Angles	
		α_A	α_B
0.6 : 1	60	29	40
1 : 1	45	28	37
1.5 : 1	33.8	26	35
2 : 1	26.6	25	35
3 : 1	18.3	25	35
5 : 1	11.3	25	37

Procedure:

1. From given information i.e. slope angle and height Draw the slope with some suitable given scale.

2. From toe of slip circle and the slope draw direction lines at angles αA and αB to intersect each other at O.

3. With respect to A as origin locate C (4.5H, H).

4. Join CO and extend it, which is locus of Centre of slip circle.

5. Choose any point on the locus. Draw trial slip circle passing through toe and find FOS for it.

6. Repeat the process for five to six trial centers along the locus of center of slip circle.

7. Plot all these FOS normal to locus line and join them by smooth curve. Find the lowest point on it which correspond to minimum factor of safety and its slip circle will be critical slip circle.

2. Friction Circle Method:

- The friction circle method, originally proposed by Taylor (1937), considers the stability of the entire sliding mass as a whole. The disadvantage of the friction circle method is that it can only be applied to a homogeneous slope with a given angle of internal friction.

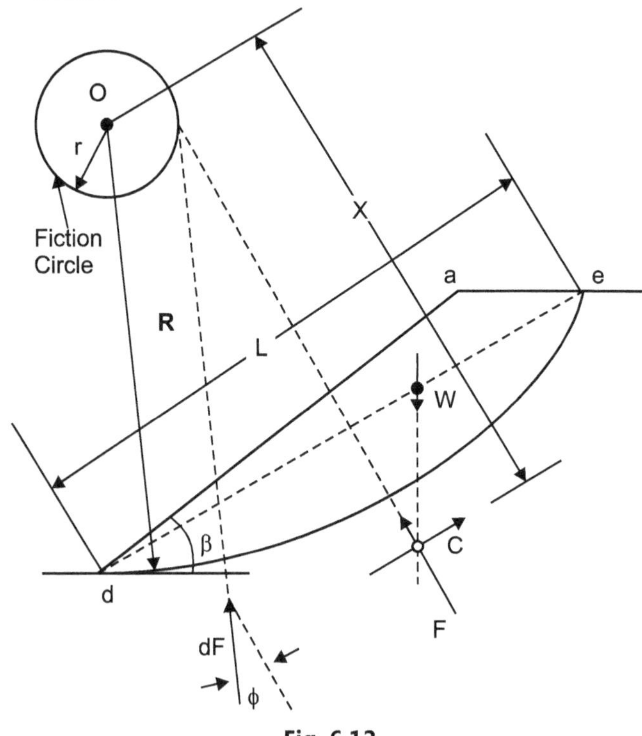

Fig. 6.12

- The forces acting on the sliding mass are its weight W, the resultant cohesion C, and the resultant F of the normal and frictional forces acting along the surface of sliding.

- Weight W acts vertically downward through centroid of sliding mass.

- Resultant cohesion C acts in a direction parallel to the chord 'd-e' and is equal to the unit cohesion c multiplied by length L of chord. Position of total cohesion C from centre of rotation is determined by

$$Cx = c \text{ (arc length 'd-e') } R$$

$$cLx = c\hat{L}R$$

i.e. $$x = R\frac{\hat{L}}{L} = \text{radius} \times \frac{\text{Arc length}}{\text{Chord length}}$$

- Sliding wedge is in equilibrium under the action of three forces so these forces should be concurrent i.e. W, C and F should pass through single point. Thus by constructing force polygon magnitude and direction of F can be determined.

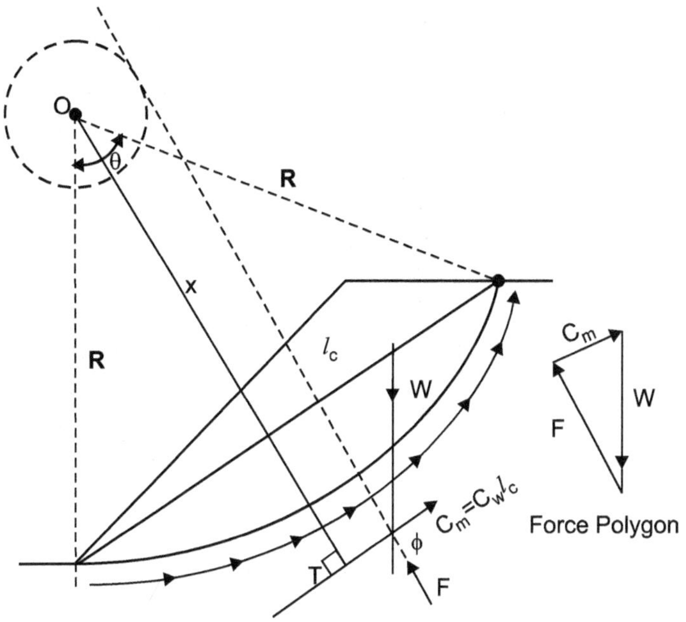

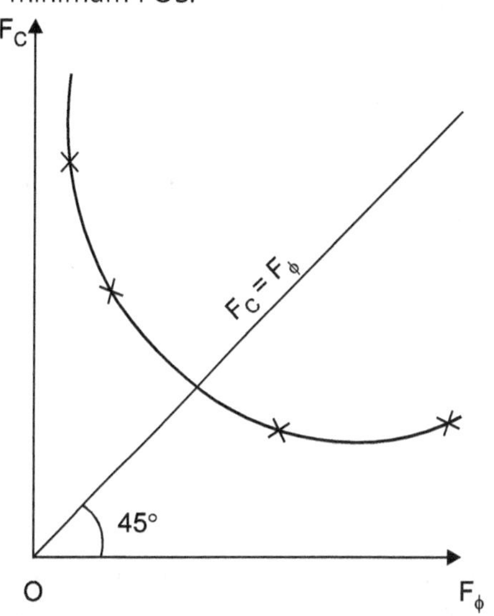

Fig. 6.13

- If factor of safety against sliding is unity, then slope is on the verge of failure. Under this condition each of elementary reaction dF must be inclined at an angle φ to the normal to the slip circle.

- As a consequence, the line of action of each elementary reaction is tangential to the circle, known as friction circle whose radius is (R sin φ) with center as center of slip circle.

Factor of Safety in Case of Friction Circle Method :

- Assume trial value of factor of safety with respect to friction F_φ.

- Calculate mobilized friction angle φ_m by using

$$\tan \varphi_m = \frac{\tan \varphi}{F_\varphi}$$

- Construct the given slope with suitable scale and trial slip circle of radius R.

- Construct the friction circle with same centre and radius r = R sin φ_m.

- Draw vertical line through the centroid of sliding mass to represent weight W whose magnitude will be (= unit weight X c/s area of sliding wedge X 1)

- Draw a line parallel to chord and at a distance x

$$\left[x = R \frac{\hat{L}}{L} \right]$$ from the centre of slip circle.

- Draw resultant reaction passing through the point of intersection of weight and cohesion.

- Construct force triangle. From force triangle measure the value of mobilized cohesion C_m

$$C_m = \frac{\text{Mobilised cohesion}}{\text{chord length}}$$

- Calculate factor of safety with respect to cohesion F_c

$$F_c = \frac{c}{C_m}$$

- Compare value of F_c and F_φ. If two values do not agree then procedure is repeated by choosing some other value of F_φ. Three trials will be sufficient to get $F_C = F_\varphi$ or alternatively plot the graph between these two values of factor of safety and draw a line at an angle of 45° to horizontal. Intersection of the line with the curve will be minimum FOS.

Fig. 6.14

3. Taylor's Stability Number and Stability Charts :

Taylor carried out stability analysis of large number of soil slopes with a wide range of properties such as slope angle (β), height (H), cohesion (c), friction angle (φ), and unit weight of soil (γ) using Friction circle method. Taylor presented the results of the stability analysis in the form of curves (stability chart) which gives the relationship between stability number (Sn) and slope angle (β) for various values of friction angle (φ). Stability number is defined by the relation $S_n = c/(\gamma H_c) = c/(\gamma FH)$.

$$c \ - \ \text{unit cohesion for the soil (kPa),}$$
$$\gamma \ - \ \text{unit weight of soil (kN/m}^3\text{),}$$
$$H \ - \ \text{height of slope}$$
$$F \ - \ \text{factor of safety and}$$
$$H_C \ - \ \text{critical height of slope}$$

The chart is divided into two zones, A and B. As shown in the inset for Zone A, the critical circle for steep slopes passes through the toe of the slope with the lowest point on the failure arc at the toe of the slope.

As shown in the inset for Zone B, for shallower slopes the lowest point of the critical circle is not at the toe, and three cases must be considered as follows:

Case 1: Where long dashed curves do not appear in the chart, the critical circle passes through the toe. This condition corresponds to Case 1. Stability numbers for Case 1 are given by the solid lines in the chart both when there is and when there is not a more dangerous circle that passes below the toe, i.e., the curves for Case 1 are an extension of the curves that correspond to a toe circle failure in Zone A.

In both Case 1 and Case 2 the failure circle passes through the soil below the toe of the slope. The depth ratio, D, which is a multiple of the slope height H, is used to define the depth (DH) from the top of the slope to an underlying strong material through which the failure circle does not pass.

Case 2: For shallow slope angles or small developed friction angles the critical circle may pass below the toe of the slope. This condition corresponds to Case 2 in the inset for Zone B. The values of Ns for this case are given in the chart by the long dashed curves.

Case 3: This case corresponds to the condition where there is an underlying strong layer at the elevation of the toe (D=1). This case is represented by short dashed lines in the chart.

Use of Chart :

To find factor of safety – Given {γ, β, φ, and c}

1. Assume trial value of FOS F_φ and determine mobilized friction angle φ_m. By using $\tan \varphi_m = \tan \varphi / F_\varphi$.

2. By using stability chart find stability number from value of $\varphi_m = \beta$.

3. Determine the factor of safety with respect to cohesion by using $F_c = \dfrac{c}{S_n \gamma H}$.

4. If $F_c \neq F_\varphi$ then procedure is repeated by choosing some other value of F_φ.

To Find Maximum Slope Angle Given {γ, F, φ, and c}

1. Determine mobilized friction angle φ_m by using $\tan \varphi_m = \tan \varphi / F_\varphi$.

2. Determine stability number by using $S_n = \dfrac{c}{F_\gamma H}$.

3. Using the stability chart determine slope angle β knowing S_n and φ_m.

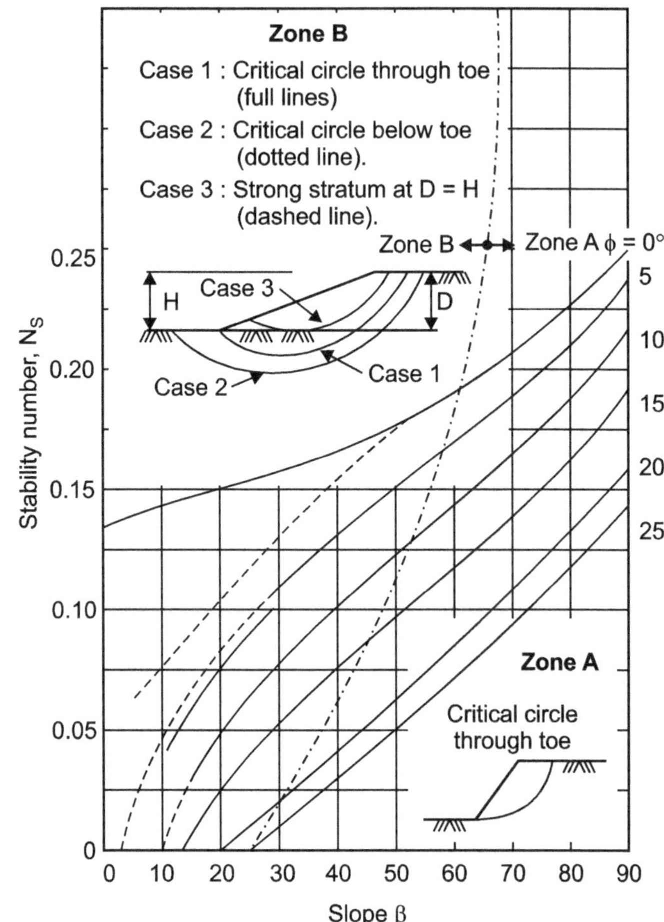

Fig. 6.15 : Taylors stability chart

Stability Conditions for Analysis

- Variations of the loads acting on slopes, and variations of shear strengths with time, result in changes in the factors of safety of slopes. As a consequence, it is often necessary to perform stability analyses corresponding to several different conditions reflecting different stages in the life of a slope. As conditions change, the factor of safety against slope instability may increase or decrease.

- When an embankment is constructed on a clay foundation, the embankment load causes the pore pressures in the foundation clay to increase. Over a period of time these excess pore pressures will dissipate, and the pore pressures will return eventually to values governed by the groundwater conditions. As the excess pore pressures dissipate, the effective stresses in the foundation clay increase, the strength of the clay will increase, and the factor of safety of the embankment will also increase the embankment height stays constant and there is no external loading, a critical condition occurs at the end of construction.

- When a slope in clay is created by excavation, the pore pressures in the clay decrease in response to the removal of the excavated material. Over time the reductions in pore pressures dissipate, and the pore pressures return eventually to values governed by the groundwater conditions. As the pore pressures increase, the effective stresses in the clay around the excavation decrease, and the factor of safety of the slope decreases with time. If the depth of excavation is constant and there are no external loads, the factor of safety decreases continually, and the minimum value is reached when the pore pressures reach equilibrium with the groundwater seepage conditions. In this case, therefore, the long-term condition is more critical than the end-of-construction condition

- In the case of a natural slope, not altered by either fill placement or excavation, there is no end-of-construction condition. The critical condition for a natural slope corresponds to whatever combination of seepage and external loading results in the lowest factor of safety. The higher the phreatic surface within the slope, and the more severe the external loading condition, the lower is the factor of safety

End-Of-Construction Stability

- Slope stability during and at the end of construction is analyzed using either drained or un-drained strengths, depending on the permeability of the soil. Undrained strength (Total stress analysis) are used for soil with low permeability and drained strength (effective stress analysis) can be used for the soil which can drain the water easily.

- For many embankment slopes the most critical condition is the end of construction. In some cases, however, there may be intermediate conditions during construction that are more critical. In some fill placement operations, including some waste fills, the fill may be placed with a slope geometry such that the stability conditions during construction are more adverse than at the end of construction.

Long-Term Stability:

- Over time after construction the soil in slopes may either swell (with increase in water content) or consolidate (with decrease in water content). Long-term stability analyses are performed to reflect the conditions after these changes have occurred.

- Shear strengths are expressed in terms of effective stresses, and the pore water pressures are estimated from the most adverse groundwater and seepage conditions anticipated during the life of the slope.

Rapid (Sudden) Drawdown:

- Rapid or sudden drawdown is caused by a lowering of the water level adjacent to a slope, at a rate so fast that the soil does not have sufficient time to drain significantly. Undrained shear strengths are assumed to apply for all but the coarsest free-draining materials ($k > 10^{-1}$ cm/sec).

- If drawdown occurs during or immediately after construction, the un-drained shear strength used in the drawdown analysis is the same as the un-drained shear strength that applies to the end-of-construction condition.

- If drawdown occurs after steady seepage conditions have developed, the un-drained strengths used in the drawdown analysis are different from those used in the end-of-construction analyses and are determined by the effective stresses during steady seepage.

- For soils that expand when wetted, the un-drained shear strength will be lower if drawdown occurs after a period of time following construction than if it occurs immediately after construction.

Earthquake:

- Earthquakes affect the stability of slopes in two ways. First, the acceleration produced by the seismic ground motion during an earthquake subjects the soil to cyclically varying forces. Second, the cyclic strains induced by the earthquake loads may result in a decrease in the shear strength of the soil.

- If the strength of the soil is reduced less than 15 percent by cyclic loading, pseudo-static analyses of the earthquake loading can be used. In pseudo-static analyses, the effect of the earthquake is represented crudely by applying a static horizontal force to the potential sliding mass

- If the strength of the soil is reduced more than 15 percent as a result of cyclic loading, dynamic analyses are needed to estimate the deformations that would result from earthquakes.

Table 6.3

Sr. No.	Analysis Case	Slope (Critical)
1.	End of construction (including staged construction)	u/s and d/s
2.	Long term (steady seepage, maximum storage pool, spillway crest or top of gates)	d/s
3.	Maximum surcharge pool	d/s
4.	Rapid drawdown	u/s
5.	Earthquake loading	u/s and d/s

SOLVED EXAMPLES

Example 6.1 : *Shear strength parameters of soil are $c = 25$ kN/m², $\varphi = 15^0$ and mobilized shear parameters $c_m = 15$kN/m² and $\varphi_m = 12^0$. Calculate the factor of safety with respect to (a) strength (b) cohesion and (c) friction. The average inter granular pressure on the failure surface is 100kN/m².*

Solution: Shear strength

$$S = c + \sigma \tan \varphi = 25 + 100 \times \tan 15$$
$$= 51.79 \text{ kPa}$$

Mobilized shear stress

$$\tau = c_m + \sigma \tan \varphi_m = 15 + 100 \times \tan 12$$
$$= 36.25 \text{ kPa}$$

Factor of safety with respect to

(a) Strength　$F_s = \dfrac{s}{\tau} = \dfrac{51.79}{36.25} = 1.43$

(b) Cohesion　$F_c = \dfrac{c}{c_m} = \dfrac{25}{15} = 1.67$

(c) Friction　$F_\varphi = \dfrac{\tan \varphi}{\tan \varphi_m} = \dfrac{\tan 15}{\tan 12} = 1.26$

Example 6.2 : *A granular soil has a saturated unit weight of 18.0 kN/m³ and an effective angle of shearing resistance of 30°. A slope is to be made of this material. If the factor of safety is to be 1.25, determine the safe angle of the slope (i) when the slope is dry or submerged and (ii) if seepage occurs at and parallel to the surface of the slope.*

Solution:

(i) When the slope is dry or submerged $F = \dfrac{\tan \varphi}{\tan \beta}$

i.e. $1.25 = \dfrac{\tan 30}{\tan \beta}$ thus $\beta = 25^0$

(ii) When seepage occurs parallel to surface of slope

$$F = \dfrac{\rho' \tan \varphi}{\gamma_{sat} \tan \beta}$$

i.e.　$1.25 = \dfrac{(18 - 9.81)}{18 \tan \beta} \tan 30$

thus　$\beta = 12^0$

Example 6.3 : *Fig. 6.16 shows an infinite slope, inclined at an angle of 25° to the horizontal. The slope is underlain by solid rock. Calculate the factor of safety, when the water table is:*

(i) at the ground surface ($S\gamma = 1$)

(ii) at 2 m below the ground surface

(iii) non-existent ($S_\gamma = 0.46$)

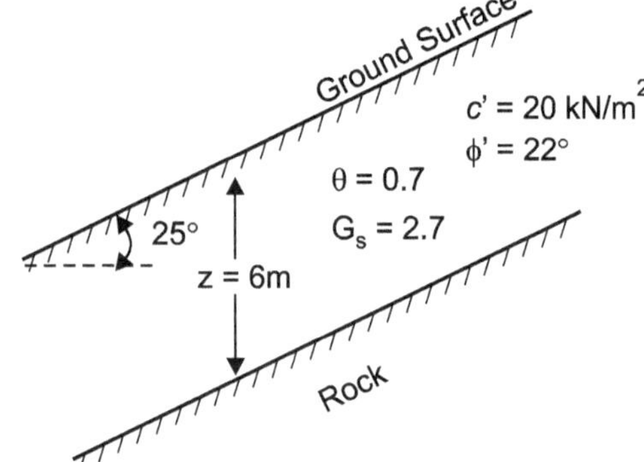

Fig. 6.16

Solution: From given soil properties unit weights of soil are as below

$$\gamma = 17.4 \text{ kN/m}^3 \quad \gamma_{sat} = 19.62 \text{ kN/m}^3$$

and　$\gamma = 9.81 \text{ kN/m}^3$

(i) Water at ground surface

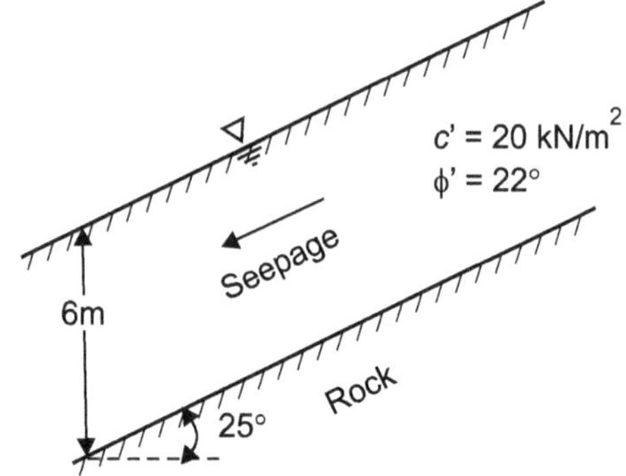

Fig. 6.17

$$F = \dfrac{c + \gamma' h \cos^2\beta \tan \varphi}{\gamma_{sat} h \sin \beta \cos \beta}$$

$$= \dfrac{20 + 9.81 \times 6 \times \cos^2 25 \times \tan 22}{19.62 \times 6 \times \sin 25 \cos 25}$$

$$= 0.88 < 1 \text{ (unstable)}$$

(ii) Water at 2 m below the ground surface

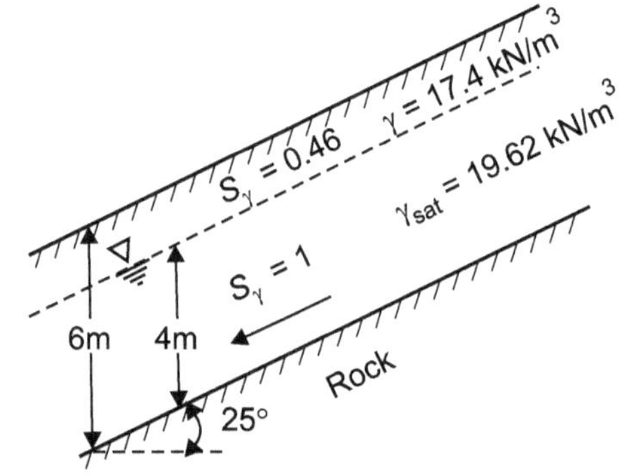

Fig. 6.18

$$F = \dfrac{c + \gamma h \cos^2 \beta \, [(h - z)\, \gamma + z\gamma'] \tan \varphi}{\gamma \sin \beta \cos \beta \, [(h - z)\, \gamma + z\gamma_{sat}]}$$

$$= \dfrac{20 + 6 \cos^2 25 \, [(6 - 4)\, 17.4 + 4 \times 9.81] \tan 22}{\sin 25 \cos 25 \, [(6 - 4)\, 17.4 + 6 \times 19.62]}$$

$$= 1.02$$

(iii) Water table is far below

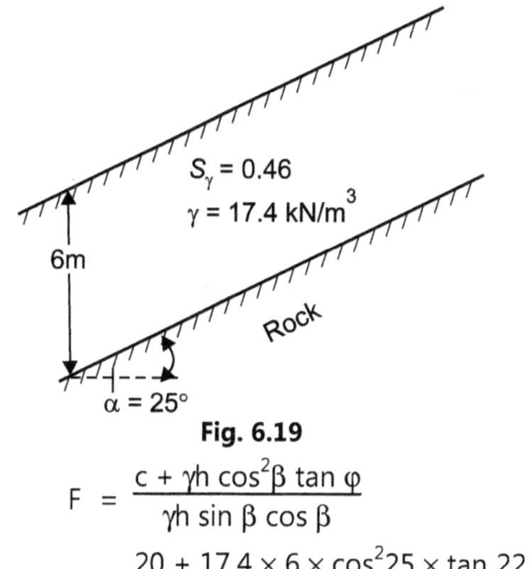

Fig. 6.19

$$F = \frac{c + \gamma h \cos^2\beta \tan\varphi}{\gamma h \sin\beta \cos\beta}$$

$$= \frac{20 + 17.4 \times 6 \times \cos^2 25 \times \tan 22}{17.4 \times 6 \times \sin 25 \cos 25}$$

$$= 1.37$$

Example 6.4 : *A 45° cutting slope is excavated to a depth of 8 m in a deep layer of saturated clay of unit weight 19 kN/m³: the relevant shear strength parameters are C= 65 kPa. Determine the factor of safety for the trial failure surface specified in Fig. 6.20.*

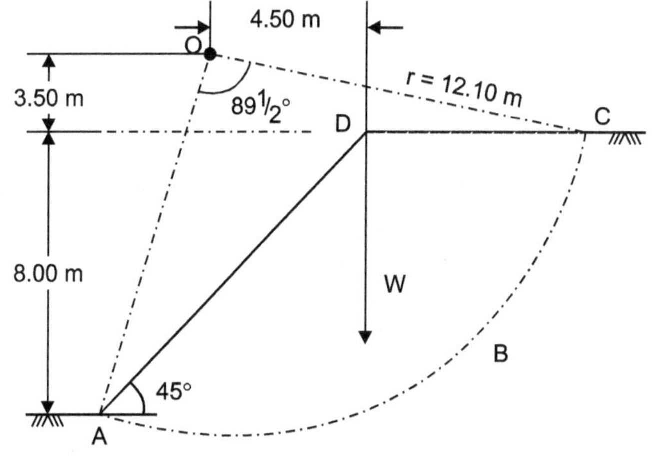

Fig. 6.20

Solution:

c/s area of ABCD (70m²) either measured with planimeter or calculated by geometry

Length of arc (=18.9m) measured from figure/ or calculated

Weight of the sliding soil mass,

$$W = \text{unit weight} \times \text{area} \times \text{thickness}$$
$$= 19 \times 70$$
$$= 1330 \text{kN/m} \text{ (taking thickness = 1m)}$$

Factor of safety $F = \dfrac{cR^2\theta}{Wd}$

$$= \frac{65 \times 12^2}{1330 \times 4.5} \times \frac{89.5}{180} = 2.48$$

Example 6.5 : *Fig. 6.21 gives details of an embankment to be made of cohesive soil with c = 20 kPa. The unit weight of the soil is 19 kN/m³. For the trial circle shown, determine the factor of safety against sliding soon after construction. The weight of the sliding sector is 329 kN acting at an eccentricity of 4.8 m from the centre of rotation. What would the factor of safety be if the shaded portion of the embankment were removed? In both cases assume that no tension crack develops.*

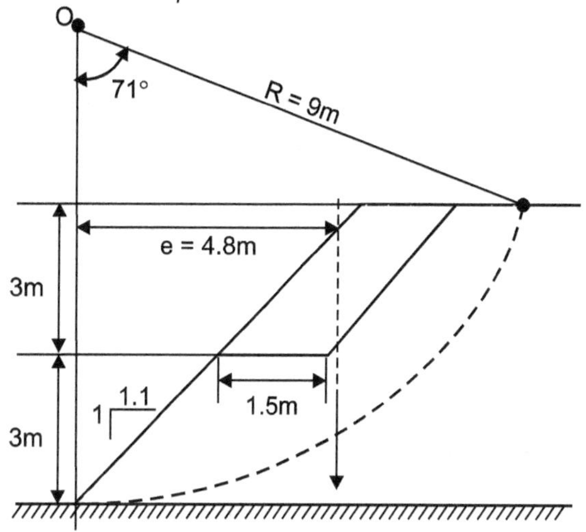

Fig. 6.21

Solution:

(i) For whole embankment (before removal of hatched portion)

$$F = \frac{cR^2\theta}{Wd} = \frac{20 \times 9^2}{329 \times 4.8} \times \frac{71}{180} = 1.27$$

(ii) For embankment (after removal of hatched portion)

Area of portion removed = $1.5 \times 3 = 4.5$m²

Weight of portion removed = $4.5 \times 19 = 85.5$kN

Position of centroid of removed portion from centre
$$= 3.3 + 0.5(3.3+1.5) = 5.7\text{m}$$

$$F = \frac{cR^2\theta}{w_1 d_1 - w_2 d_2}$$

$$= \frac{20 \times 9^2}{329 \times 4.8 - 85.5 \times 5.7} \times \frac{71}{180} = 1.84$$

Example 6.6 : *Fig. 6.22 shows 10m high slope with soil properties r = 19kN/m³ and c = 70kPa. Determine factor of safety by using method of slices for the trial circle as shown.*

Solution: Draw the slope with failure surface with scale and then divide the sliding soil in to (any) number of slices as shown in figure (nine number). Draw mid-ordinate for each slice, measure width of each slice, mid-ordinate of slice and angle made by bottom of mid- ordinate with radial line. Enter all these values in a tabular column as shown below. Calculate weight of each slice (w = r x b x h). Calculation are shown in the tabular column. Calculate factor of safety by using

$$F = \frac{c\hat{L} + \sum N \tan\varphi}{\sum T} = \frac{20 \times 23.48 + 1307.21 \times \tan 20}{724.2}$$

$$= 1.305$$

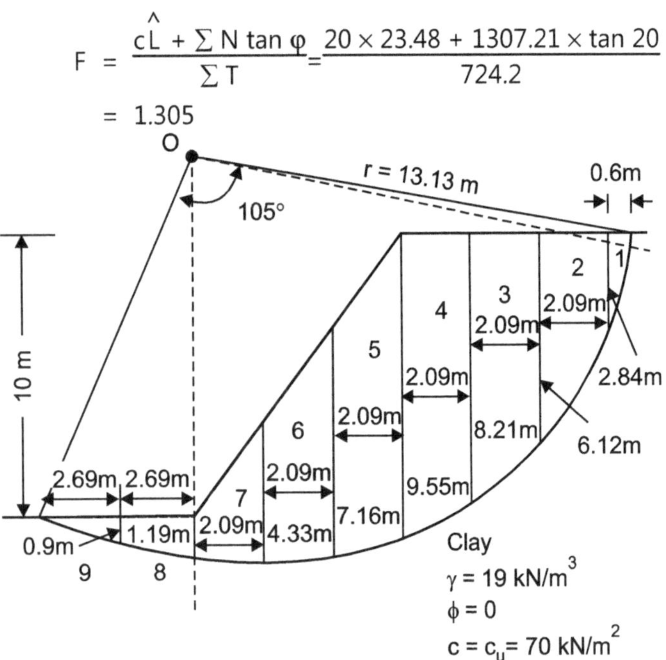

Fig. 6.22

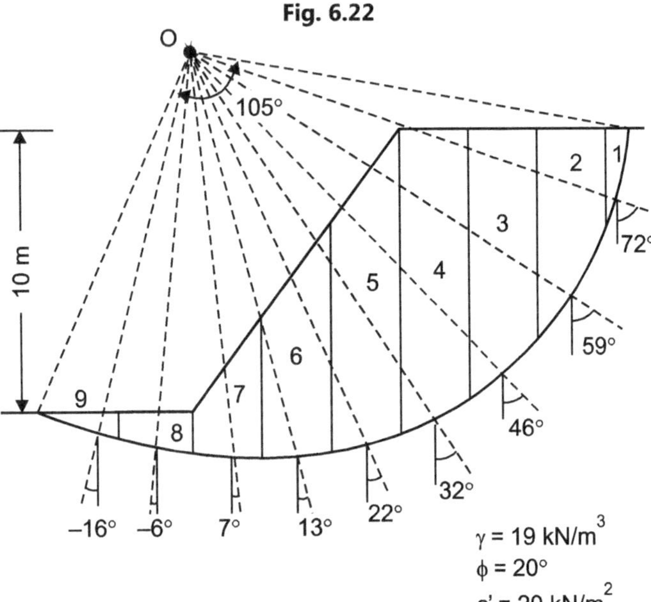

Fig. 6.23

Table 6.4

Slice No. (1)	W_n (kN/m) (2)	α_n (deg) (3)	$\sin\alpha_n$ (4)	$\cos\alpha_n$ (5)	ΔL_n (6)	$W_n \sin\alpha_n$ (7)	$W_n \cos\alpha_n$ (kN/m) (8)
1.	16.188	72	0.951	0.309	1.942	15.395	5.00
2.	177.9	59	0.788	0.515	4.058	140.185	91.62
3.	284.52	46	0.719	0.695	3.007	204.57	197.74
4.	352.62	32	0.530	0.848	2.465	186.89	299.02
5.	331.78	22	0.375	0.927	2.255	124.42	307.56
6.	228.13	13	0.225	0.974	2.146	51.33	222.2
7.	109.6	7	0.122	0.993	2.105	13.37	108.83
8.	53.41	−6	−0.105	0.995	2.704	−5.61	53.14
9.	23.0	−16	−0.276	0.961	2.799	−6.35	22.10
Note : $\Delta L_n = \dfrac{b_n}{\cos\alpha_n}$					$\sum = $ 23.48 m	$\sum = 724.2$ kN/m	$\sum = 1307.21$ kN/m

1. What are the assumptions that are generally made in the analysis of the stability of slopes ? Discuss their validity briefly.
2. What are the different types of slope failure ?
3. Differentiate between finite and infinite slopes.
4. What is a stability number ? What is its utility in the analysis of stability of slopes ?
5. What are the factors that affect the stability of slope ?
6. Draw neat sketches to show :
   (i) Failure of a finite slope.
   (ii) Toe failure
   (iii) Base failure.

1. A simple homogeneous slope has a stability number of 0.16. What would be its safe height to allow a factor of safety of 1.5, if the soil has C = 25 kN/m² and γ = 20 kN/m³ ?
2. A slope having inclination of 30° with the horizontal is to be constructed with soil having following properties : C = 15 kN/m², γ = 19 kN/m³ and φ = 22°. Determine the safe height if the factor of safety is to be 1.50.
3. A long natural slope in an over-consolidated clay (C' = 10 kN/m², φ = 25°, γ_{sat} = 20 kN/m³) is inclined at 10° to the horizontal. The water table is at the surface and the seepage is parallel to the slope. If a plane slip has developed at a depth of 5 m below the surface, determine the factor of safety.

   (**Ans.** F_s = 1.90)
4. A vertical cut is to be made in clayey soil for which tests gave C = 30 kN/m², γ = 16 kN/m³ and φ = 0. Find the maximum height for which the cut may be temporarily unsupported. For φ = 0, i = 90°, the value of stability number is 0.261.

   (**Ans.** 7.18 m)
5. If the stability number for each slope 10 m high is 0.056, determine its factor of safety given φ = 20°, C = 30 kN/m², γ = 10 kN/m³. (**Ans.** F_s = 2.68)
6. A 40° slope is excavated to a depth of 8 m in a deep layer of saturated clay. Determine the factor of safety of the slope, if the soil has C = 20 kN/m², φ = 150° and γ = 20 kN/m³.

MODEL QUESTION PAPERS FOR
End-Semester Examination

Paper- I

Time : 3 Hrs.   **Marks : 60**

Instructions to the candidates :

(1) Each Question carries 12 Marks.

(2) Attempt any five questions from the following.

(3) Illustrate your answers with neat sketches, diagram etc., wherever necessary.

(4) If some part or parameter is noticed to be missing, you may appropriately assume it and should mention it clearly.

1. (a) Explain with sketches the terms   **[6]**

    (i)   Inside clearance.

    (ii) Outside clearance.

    (iii) Area ratio.

    Comment upon its usefulness.

(b) A multi-storeyed building is to be constructed on a bank of a river, when rock bed is expected at about 9 m depth. What will be your plan of action for geotechnical investigation ?   **[6]**

2. (a) Explain Skempton's analysis of determination of bearing capacity of Clayey soil.   **[6]**

(b) Determine the ultimate and net bearing capacity, use following data: **[6]**

    (i)   Footing size = 2 m $\times$ 2 m, depth of foundation = 1.5 m.

    (ii) Soil density = 1800 kg/m^3 , c = 15 kN/m^2, ϕ = 15°.

    (iii) N_c' = 9.7, N_q' = 2.7, N_γ' = 0.9.

    (iv) Use local shear concept.

3. (a) What are the problems associated with expansive soil?   **[6]**

(b) What are the characteristics of expansive soil?   **[6]**

4. (a) How do you calculate the bearing capacity of raft ?   **[6]**

(b) How you determine contact pressure for a footing subjected to eccentric loads ?   **[6]**

5. (a) State and explain five different methods of anchorage of sheet piles.   **[6]**

(b) A group of piles consists of 15 piles arranged in three rows and five columns. Compute efficiency of pile group by following method. Assume diameter of piles = 300 mm spacing 0.75 m centre to centre.   **[6]**

    1.   Converse – Labbare's formula

    2.   Los-Angles formula

    3.   Feld's rule.

6. (a) What are the factors that affect the stability of slope ?   **[6]**

(b) A slope having inclination of 30° with the horizontal is to be constructed with soil having following properties :
C = 15 kN/m^2, γ = 19 kN/m^3 and ϕ = 22°.

Determine the safe height if the factor of safety is to be 1.50.   **[6]**

Paper- II

Time : 3 Hrs.   **Marks : 60**

Instructions to the candidates :

(1) Each Question carries 12 Marks.

(2) Attempt any five questions from the following.

(3) Illustrate your answers with neat sketches, diagram etc., wherever necessary.

(4) If some part or parameter is noticed to be missing, you may appropriately assume it and should mention it clearly.

---

1. **(a)** At depth of 4 m, blow count was 18. If the ground is saturated, find corrected value of N if unit weight of sound is 1.9 gm/cm^3. **[6]**

(b) Draw a neat sketch of SCPT and explain how this test is performed in the field. **[6]**

2. **(a)** Enlist the causes of differential settlement and explain how to minimise it. **[6]**

(b) Determine the safe bearing capacity of a circular footing for the following details : **[6]**

Diameter = 2m, Depth of foundation = 1.5 m

Shear strength of soil (S_u) = 25 kN/m^2. Unit weight of soil = 20 kN/m^3

Value of N_c = 5.7, N_q = 1.0, N_γ = 0.0 Factor of safety = 3.0

3. **(a)** Can we reduce the damage to structure resting on difficult soil by using geo-synthetics. **[6]**

(b) Explain the spatial distribution of expansive soil in India. **[6]**

4. **(a)** Design a footing to support two columns at c/c spacing of 6m carrying load of 3000kN and 4500kN. Lighter column is near the boundary (0.3m from edge of column) assume column size 0.5m × 0.5m and SBC = 350kPa. **[6]**

(b) What is raft foundation? When it is preferred ? **[6]**

5. **(a)** Draw the sketches of structural arrangements involved in : **[6]**

    (i)   Cantilever sheet pile.

    (ii) Anchored sheet pile.

(b) Explain how do you decide bearing capacity of single pile by any one of the following method : **[6]**

    (i)   Static method.

    (ii) Dynamic method.

    (iii) Load test method.

6. **(a)** What is a stability number ? What is its utility in the analysis of stability of slopes ? **[6]**

(b) A 40° slope is excavated to a depth of 8 m in a deep layer of saturated clay. Determine the factor of safety of the slope, if the soil has C = 20 kN/m^2, ϕ = 150° and γ = 20 kN/m^3. **[6]**